Growth of Literature and Measures of Scientific Productivity

Scientometric Models

Growth of Literature and Measures of Scientific Productivity

Scientometric Models

I. K. RAVICHANDRA RAO
Professor and Head
Documentation Research and Training Centre
Indian Statistical Institute
8th Mile, Mysore Road
Bangalore-560059

Published By

Ess Ess Publications
New Delhi
for

Sarada Ranganathan Endowment for Library Science
Bangalore

Sarada Ranganathan Endowment Lectures (27) (2008)

S.R. Ranganathan's
Five Laws of Library Science

Books are for Use
Every Reader his/her Book
Every Book its Reader
Save the Time of the Reader
Library is a Growing Organism

ISBN: 978-81-7000-609-1

Ess Ess
Rs.450/-

First Published 2010

Published by:
Ess Ess Publications
4837/24, Ansari Road,
Darya Ganj,
New Delhi-110 002.
INDIA
Phones: 23260807, 41563444
Fax: 41563334
E-mail: info@essessreference.com
www.essessreference.com

for

Sarada Ranganathan Endowment for Library Science
702, 'Upstairs', 42nd Cross, III Block, Rajajinagar,
Bangalore-560 010
E-mail: srels@dataone.in Phone: 080-23305109

Printed and bound in India at Salasar Imaging Systems.

CONTENTS

FOREWORD

Professor Neelameghan, Professor Ravichandra Rao, Professor Raghavan and friends: I feel deeply honoured to be asked to preside over today's first lecture in a series by Professor Rao and to inaugurate this series under the Sarada Ranganathan Endowment for Library Science. But I also feel somewhat out of place in this gathering as I do not belong to this field. It is also my disappointment that Dr K N Prasad, who invited me in the first place and gave me all the details, is unable to be present. Let me offer some remarks on scientometrics from the viewpoint of a physicist who is now a senior citizen.

Scientometrics is a relatively recent development, a little over thirty years old. I have seen the outlines and the topics that Professor Rao plans to present – growth and obsolescence of literature; measures of scientific productivity; and the nature of collaborative work in science. He has also mentioned in his abstract that as a field scientometrics is unpopular among some scientists at least, so he will elaborate on that. I thought of saying something about that aspect and why it is so, clearly a delicate balancing act is needed.

As you all know, science in the modern sense is only about 400 years old, and the practice of publishing in professional journals is a bit younger. I would say that even as recently as about 50 years ago, the main source of creative scientific work was an individual, or sometimes a group of two or three collaborators. Certainly in the 1920's, 30's and 40's when the great modern theories of physics were created, the active community was quite small. In a specialized area, every scientist was essentially personally familiar with others in that area. Ideas and discoveries were often communicated first by word of mouth or even by postcards.

However soon after World War II, certainly in physics and in particular in some parts of experimental physics, it became common for groups of 10 or 15 physicists to work together on an experiment. As experimental facilities and investments grew, these groups became large too – first 50, then 100 and now in some cases several hundred. Of course I refer here to very large scale experiments. At one point, when a paper was published by such a large group in a journal, the editors would not list the names of all the authors but just referred to them by some code name – like a football or cricket team. At the same time the community of physicists also increased enormously, and it became more and more difficult for one scientist to be in touch with all others in even a specialized area.

Scientific creativity and research have often been compared to creativity in the humanities such as literature and music. It has for so long been centred on the individual, and the joy that one experiences when an important discovery is made is quite similar to that in art. So one can understand that when in some fields it became unavoidable to have larger and larger collaborations, we felt we were going to lose something precious about the creative experience in science. I would like to quote some sentences from the writings of a very great physicist of the

20th century, Eugene Paul Wigner, on this aspect. Basically arguing for protecting the individual and giving value and esteem to 'little science' as against 'big science', he says:

'One does not have the satisfaction which creative work, as we know it today, provides, if one's activities are too closely directed by others.'

'It is hard to imagine how deep insights can be developed other than in comparative solitude.'

He also said something about the joy of doing science:

'It has been said that the only occupations which bring true joy and satisfaction are those of poets, artists and scientists, and of these the scientists are apparently the happiest.'

So I think you can at least appreciate why scientists of my generation, for example, feel concerned and a little anxious with the trends to measure, quantify and calculate scientific productivity in so many ways. Would you like to do this in music, painting or literature? It is true that public investments in science have gone up, very large institutions have been created, and at some stage for some purposes one needs to have reliable estimates of returns on investments, as for example in agriculture or industrial or pharmaceutical production. Still, is there some way to protect the role of the individual in scientific creativity and not let it drown in numbers, indices and measures?

As an aside I might mention that scientometric assessment of individuals has in some ways led to unhealthy competition and even to a rise in plagiarism.

You may be aware of a journal called 'Current Science' published here in Bangalore. It was started many years ago in the early 1930's, and is something the entire Indian scientific community reads regularly. The Chief

Editor is Professor P. Balaram, also Director of the Indian Institute of Science, and over many years his elegant editorials have become justifiably famous. In the August 25, 2008 issue he wrote an editorial titled 'Scientometrics – A Dismal Science'. I quote parts of it to convince you that the concerns I have described are genuine:

'The world of science has never been the same since Eugene Garfield introduced the Science Citation Index and the idea of the Journal Impact Factor. Slow to be accepted at the beginning, these tools have now been sharpened by the information technology revolution; ensuring that scientists and administrators, editors and publishers self consciously examine performance indices at every step. The joy of science seems to have disappeared for many individuals, who are now consumed by impact factors and citation analysis. There is a more calculated and purposeful approach to science today; the refreshing naivete of the amateur has been replaced by the, sometimes, disturbing realism of the professional manager.'

He also quotes from an essay titled 'Escape from the impact factor' by Philip Campbell, editor of Nature:

'I am sure that citation statistics of large numbers of individual papers can reflect the impact of contributions at the institutional or national level. They can also provide useful input at the level of an individual. But for a sure assessment of an individual, there is truly no substitute for reading the papers themselves, regardless of the journals in which they appear'

And Balaram also brings in a sentence from Shakespeare's 'Julius Caesar':

'Are all thy conquests, glories, triumphs, spoils Shrunk to this little measure?'

Let me now conclude. With the increased scale and expectations of science, vastly greater funding and involvement in public policy, the field of scientometrics may be necessary for policy makers and governments to make important decisions. In developing this new discipline and teaching it to librarians and information professionals, you might like to see how it is viewed by different groups of scientists themselves: how do the views of physicists, mathematicians, chemists and biologists differ from one another? What about the older and the younger generations, what about the views of social scientists? And foremost – how can the role of the individual, the joy of scientific discovery, the joy of teaching and training the next generation, all be retained while using scientometric methods, data, indices for those purposes for which they are useful and even essential?

I hope these thoughts will be in the background as Professor Rao develops the technicalities of the subject in his lectures.

Thank you very much again for inviting me to this inaugural lecture.

N. Mukunda

Professor (Retd.), Indian Institute of Science, Bangalore and Editor, Publications, Indian Academy of Sciences, Bangalore

Preface

Design and development of science indicators are now increasingly becoming popular. Numbers of papers, growth rates (of articles, authors, patents, journals, institutions, etc.), impact factors, h-index, etc. are simple indicators. Computations of these indicators are not simple. It requires large data from several databases; it involves high cost and time; the other factors such as reliability, validity and availability of data are equally important. Further, it is also difficult to identify and select reasonably good methodologies to analyze scientometric data. Analysis of scientometric data helps us in

- Formulating science and technology policy;
- Identifying the relative position of research communities;
- Developing measures to evaluate R&D; and
- Studying the growth of literature in a given field.

Generally, the science indicators are of two types – input indicators and output indicators. The input indicators are derived from data on human and material resources. The output indicators are classified as quantitative and qualitative types. Counting of publications, citations, information products, patents, performance output (on time and on budget) etc. are qualitative output indicators. Judgment evaluation, goal achievement, impact on user satisfaction, etc. are the qualitative output indicators. Keeping these factors in view, in this lecture series, I shall be discussing mainly

- Different growth models as applicable to growth of literature;
- Indicators related to scientific productivity of scientists; and
- Issues in scientometrics.

The literature in scientometrics indicates that a power model fits best in cases of convex growth and that a Gompertz model generally fits in cases of S-shaped growth. In modeling a growth of literature pattern, the choice between an exponential and a logical function may lead to false conclusion. These issues together with various characteristics and selection of growth models are discussed. Among the various studies in scientometrics, author productivity studies are most common. Until the 1970s, most of the studies were concentrated on Lotka's law. In recent times, however, the research is mainly centered on impact factors, h-index and other citation measures. These topics along with some issues in scientometrics are also discussed.

It is hoped that this document will be useful to students as well as researchers in scientometrics.

Date: 9th August 2009. **I K RAVICHANDRA RAO**

Sarada Ranganathan Endowment for Library Science (SRELS), Bangalore

The Sarada Ranganathan Endowment for Library Science (SRELS), founded by Dr. S.R. Ranganathan in 1961, is registered as a Charitable Endowment with the Government of India (1963).

The aims of the Endowment briefly are:

- To improve the library and information services in India
- To train library and information service personnel
- To promote the provision of efficient library and information service
- To apply research results to library and information studies

The objectives of the Endowment are:

- Furthering the cause of library and information service
- Organising seminars and workshops on recent developments in library and information science and related fields.
- Organising periodically courses of lectures on recent developments in library and information science and related fields by national and international resource persons.
- Supporting the publication of these lectures and of other nascent ideas in the field of library and information science.

There are three series of lectures organized under the auspices of the Endowment:

1. Sarada Ranganathan Endowment Lectures. (1965-);

2. Dr. S.R. Ranganathan Memorial Lectures. (1973-); and
3. Curzon-Seshachalam Endowment Lectures. (2001-).

The Endowment publishes a periodical entitled ***SRELS Journal of Information Management*** (formerly, ***Library Science with a slant to Documentation*** (from 1964 to 1988), and ***Library Science with a slant to Documentation and Information Studies*** (from 1988 to 1999)) and the texts of the series of lectures.

Growth of Literature and Measures of Scientific Productivity: Scientometric Models is the text of the 27th Sarada Ranganathan Endowment Lectures delivered by Dr. I.K. Ravichandra Rao, Professor and Head, Documentation Research and Training Centre, Indian Statistical Institute, during 18-20 December 2008 in Bangalore.

The lectures cover two types of science indication – quantitative and qualitative. Different growth models applicable to growth of literature; Indicators related to scientific productivity of scientists including an their productivity; and Issues in scientometrics are discussed in the their lectures.

Prof. N. Mukunda, Professor (Retd.), Indian Institute of Science, Bangalore and Editor, Publications, Indian Academy of Sciences, Bangalore, while inaugurating the lecture series, offered some remarks on scientometrics from the viewpoint of a physicist. He reiterates that with the increased scale and expectations of science, the field of scientometrics may become necessary for policy makers in governments to make important decisions.

K.N. Prasad
Executive Officer
SRELS

Acknowledgement

I am grateful to the Board of Trustees of the Sarada Ranganathan Endowment for Library Science (SRELS) for giving me an opportunity to deliver the Sarada Ranganathan Endowment Lectures (27) (2008). I also thank Professor A Neelameghan, SRELS and Professor K S Raghavan, Documentation research and Training Centre, Indian Statistical Institute, Bangalore, for their valuable suggestions in preparing this document. I am also thankful to Dr K N Prasad, Executive Officer, SRELS, for all the support he has given me.

I K RAVICHANDRA RAO
Professor and Head
Documentation Research and Training Centre
Indian Statistical Institute

About the Author

RAVICHANDRA RAO, INNA KEDAGE is on the Faculty of the Documentation Research and Training Centre (DRTC) of the Indian Statistical Institute (ISI). Currently he is Professor and Head of DRTC. He was Head of the Indian Statistical Institute, Bangalore Centre from Oct. 2002 to Sept. 2007.

After having completed BSc. in Mathematics and Statistics from the Mahatma Gandhi Memorial College, Udupi in 1968, Prof. Rao obtained his Master's degree in Statistics and Diploma in Computer Science from the Indian Statistical Institute, Calcutta. In 1975, he did Associateship in Documentation from DRTC. He received his Ph.D. degree from the University of Western Ontario, Canada, in 1981.

Prof. Ravichandra Rao joined DRTC in 1970. Since then he is involved in teaching and research in applied statistics, (bibliometrics/informetrics and scientometrics), programming, library automation, information retrieval, information management and DBMS with special emphasis on bibliographical databases. He has been on the UNESCO/UNDP, IDRC missions following :

- For evaluation of Village Information Centers, organized by the Manipal Industrial Trust, which was sponsored by the IDRC, Canada, in 1992.

- In Quantitative Methods for Library and Information Science, to design curriculum of PG course in Information Science, for the Latin American Region at Caracas, Venezuela, in June 1984.
- To deliver a course of lectures in Bibliometry and Scientometry in 1981, to the students of the 4th PG course in formation science sponsored by the UNESCO at the Institute of Library Science of the University of Philippines, Manila.

He worked as Visiting Professor at Limburgs University (now known as University of Hasselt), Diepenbeek, Belgium in 1990, 2001, 2007 and at Addis Ababa University, Ehiopia in 1990. He has visited a number of library schools in India as a resource person to deliver a series of special lectures. Prof. Rao is actively involved in research in the area of informetrics and scientometrics and is associated with the international conferences organized by the ISSI, The Netherlands. He was a founder Executive Committee Member of the International Society for Scientometrics and Informetrics, The Netherlands (up to 2005). His research interests include growth and obsolescence of literature, library statistics, theories of bibliometrics, scientometrics and informetrics; Library automation information retrieval, information management, in addition to the development of software for library applications. Based on his research and teaching experience, he has authored two books (Quantitative Methods in Library and Information Science, (it has been translated into Portuguese language.) and Library Automation) Both the books have been recognized as textbooks for students at the post-graduate level in some of the Indian and foreign universities. He has edited a number of seminar volumes both at the national and international levels and also was a Guest Editor for a few journals in Information field in India. He has more than sixty research papers to his credit and has reviewed a number of books and articles.

He was awarded the Fellowship of the Society for Information Science, New Delhi, in 1996 for his contribution to Information Science.

Introduction

P. C. Mahalanobis, founder of the Indian Statistical Institute, said in the 1950s that statistics is a "key technology" in all development and forecasting studies. Statistics is applied to a number of areas such as perspective planning, industrial and agricultural development and others. Statistics and related technologies (— generally known as quantitative methods / techniques) when applied in depth to a field may give rise to a new subject - for example, econometrics and psychometry. During the early 20th century, the application of quantitative methods to library and information field was known as "statistical bibliography" (Pritchard, 1969.) Hulme[56] in 1923 traced the root of this term to its first use, in his summary of the work done in 1917 by Cole and Eales. Dr S.R.Ranganathan was one of the few researchers who applied quantitative methods to library management in 1930s. At the ASLIB Conference[5] (1948) in Leamington Spa, he presented the idea that it is necessary for librarians to develop "librametry" on the lines of biometry, econometry, and psycometry since "many of the matters connected with library work and services involve large numbers."

Pritchard[87] (1969) used the term "bibliometrics" to describe all "studies to quantify the process of written communication." Fairthorne[43] (1969) defined it as "the quantitative treatment of the properties of recorded discourse and behavior pertaining to it." Although the term librametry was well defined, the term was hardly known among the professionals in library and information field! For instance, Brookes[14] (1990) in his address at the 2nd International Conference on Informetrics held in Canada, said that "had I known of Ranganathan"s term in time, I would have adopted "librametry" for information studies and "bibliometrics" for information science. But it was too late; librarians liked bibliometrics."

In the 1960s, particularly in Eastern Europe, the term "scientometrics" was used to denote "measurement of informatics process." The term informatics was then widely used to mean "documentation / information handling activities;" obviously, there is not much difference between bibliometrics of the West and the scientometrics of the East Europe. In another development, Otto Nacke[14] of West Germany in 1979 proposed a term "informetrics." FID constituted a committee with this name and Nacke was its first Chairman. Rajan[88], the next Chairman of the Committee, reformulated the objectives of informetrics as to (1) provide reliable data for research and development, policy-making, planning; (2) to evaluate institutions, projects, articles, products, and other academic activities, and (3) to identify or to develop the techniques to trace the origins and development of concepts. In a short communication on "Informetrics vis-à-vis Bibliometrics: Scope and its Development", Ravichandra Rao[91] mentioned that it is a field wherein the flow of information and behaviour of information are analyzed, measured and quantitative relationships are established. It is a scientific field wherein the developments of measurement of impact of information are assessed continuously.

Bibliometrics may be treated as synonymous to informetrics having a scope to analyze quantitative characteristics of information.

The fact is that whether we call our research area as librametry or bibliometry or scientometrics or informetrics most of the topics we deal with are:

- Quantitative aspects of library and information studies, especially use and user studies, growth of collection, age distribution of documents, circulation statistics, etc.
- Journal productivity (by coverage, by use, by citation, cost-effectiveness measures, impact factor, h-index, sources of citations, immediacy of citations, age of sources cited, coverage in databases etc.)
- Measures of productivity or author productivity, including studies related to multiple authorship (number of publications. cost-effectiveness measures, impact factor, h-index, reprints request, photo copies made, sources of citations, immediacy of citations, number of reviews, adoption rates (text books), etc.)
- Obsolescence and growth of literature
- Co-citation, bibliographic coupling, co-word analysis, rank distribution of words, etc.
- Quantitative analyses of science (science indicators — country-wise, language-wise, subject-wise, etc.)
- Identifying relationships among various disciplines, structure of subjects, etc.
- Evaluation of scientific research (by institutions, by individuals, by countries, etc.)

Acceptance of a single term to define a subject and acceptance of its scope are necessary for any scientist. Otherwise, it is difficult to include it in a syllabus. It is also difficult to get research grant from different agencies. It helps us in identifying the research groups especially at national and international levels. In this lecture series, the term **scientometrics** is used and all other terms such as librametry, bibliometrics and informetrics as synonymous with informetrics. Further, these lectures will mostly cover, 'growth of literature' (including some aspects of obsolescence) and 'author productivity' (including research collaboration)

Lecture 1

GROWTH OF LITERATURE

The number of scientific peeriodicals including abstracting periodicals are simple indicators of scientific growth. Price[84] in 1961 argued that scientific literature grows exponentially and computed the growth rate as 5% over the previous two centuries. He further observed that literature doubles approximately once in fifteen years. Neelameghan[80] analyzed the documents on the history of medicine in India for the period 1954-61, during which period Indian contribution was 65% and foreign was 30%. He studied the growth of Indian medical societies and medical periodicals between 1780 and 1920. He also studied the coverage of Indian medical literature in Index Medicus and Experta Medica and it was found that they covered respectively only 38% and 13.5% of the Indian literature. Since then a number of articles have been published on this topic, particularly on the growth of literature in different subjects and on various growth models. The number and growth characteristics (of articles, periodicals, scientists, discoveries, etc.) have been matters of some debate for a considerable time. For instance, Price[84] argued that:

- Once in fifty years the number of universities, labour force, population, etc. double.
- Once in twenty years GNP, discoveries, scientists, college entrants/1000 population double
- Once in fifteen years the number of science periodicals doubles
- Once in ten years the number of articles / literature in a field (particularly in science) doubles

Once in two years the number of web sites doubles (for example, according to Netcraft web server survey, there were 108,810,358 web sites in February 2007 as against 70,396,562 web sites in August 2005. There were 185,167,897 web sites in November 20 08. Figures 1, 2 and 3 show the growth of web sites, active servers and the market shares of top servers. (Netscape survey: *http://survey.netcraft.com/Reports/200811*)

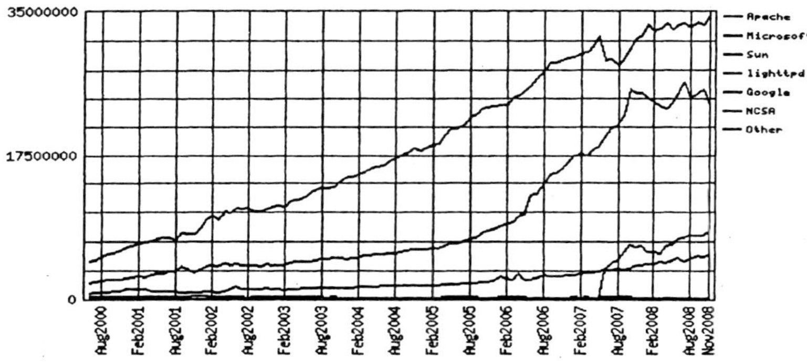

Fig. 1. Total Sites Across All Domains August 1995 - November 2008

As observed by many, the growth of publications passes through the following four stages (Price[84], Michael Mabe[70] and several others):

1. the preliminary period of growth in which the absolute increments are small although the rate of increase is high;

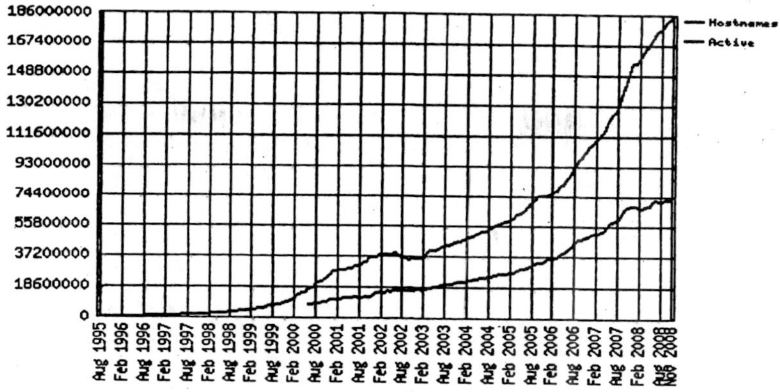

Fig. 2. Totals for Active Servers across all Domains: June 2000 - November 2008

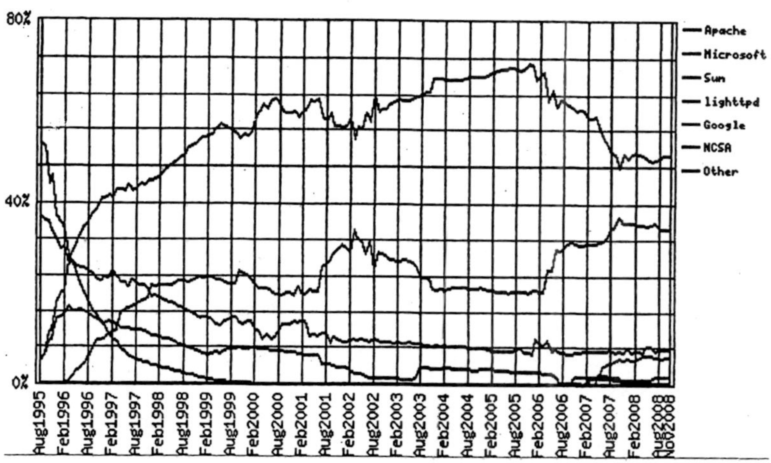

Fig. 3. Market Share for Top Servers across all Domains August 1995 - November 2008

2. the period of exponential growth when the number of publications in a field doubles at regular intervals as a result of a high rate of growth;
3. the period when the rate of growth declines but the annual increments still remain approximately constant; and
4. the final period when both the rate of increment and

the absolute increase decline and eventually approach zero.

The growth curves which explain the above four steps are also referred to as S-shape curve. A typical S-shaped curve is given in Figure 4.

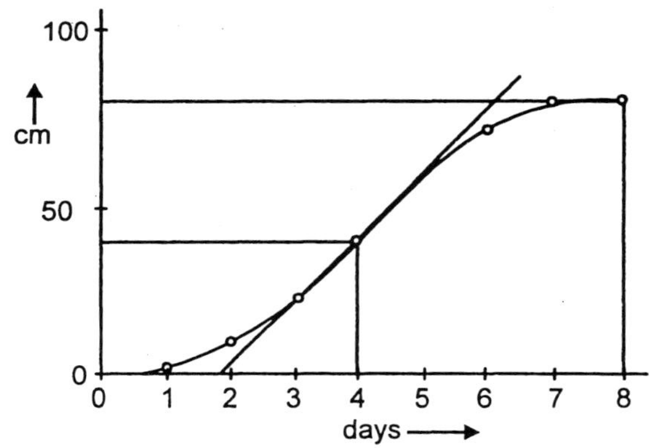

Fig. 4. A Typical S-shaped Curve
(Source: *Little Science Big Science* by Price[84])

The growth generally adopts an S-shaped pattern and is symmetrical about its point of inflection. Price in 1963 estimated that the number of scholarly periodical titles being published at the end of the twentieth century would exceed one million. Meadows (referred from Mabe[70]) on the other hand observed that estimates of the number of periodicals varied from 10,000 in 1951 to 70,000 in 1987. Price argued that the logistic growth of knowledge over a period of time is a result of a number of applications of intellectual innovations. He has fitted the logistic curve to the cumulative number of new publications appearing every year in science. He has also studied the growth of literature covered by *Physics Abstracts* during 1900-1950. He observed that except for the interruptions during the two world wars the literature has been increasing at exponential rates with a doubling time of about twelve years. Mabe in his study also observed that journal growth rates have been remarkably consistent over time with average rates of 3.46%, since 1800. He has, in fact, observed:

- From 1900 to 1940, the number of active journal titles increased at an actual rate of 3.23%, a doubling time of twenty-two years.
- From 1945 to 1976, the number of journals increased at an annual rate of 4.35%, representing a doubling time of sixteen years.
- Since 1977, the number of journals increased at 3.26%.

Growth rates during 1900 to 1940 and since 1977 are almost identical, despite the fact that the two periods are completely different in many ways. Of course, during the post world war period, the growth rates were very high; this trend continued until mid-1970s. The slow growth rates after the mid-1970s were perhaps due to

- The oil crisis of the 1970s
- The increasing public awareness of potential ecological disaster and
- The turning away from nuclear technology in the 1950s.

Further, these factors, certainly lead to a slow down of government support for research. With the following three assumptions

1. the number of periodicals in a given subject grows exponentially in time,
2. concurrently each periodical also augments the number of papers on the subject exponentially in time,
3. the rate of growth of articles in an individual periodical is the same for all periodicals,

Yamazaki studied the number and rate of growth of scientific periodicals.. His review critically assesses studies based on the 'ecological approach' to the growth of periodicals publishing. He concluded that the annual rate of growth of scientific periodicals is 1.85% from the end of the eighteenth century. Naranan[78] has shown that 'a frequency distribution (J(p)) of the number of periodicals with p articles is of the form

$$J(p) \propto p^{-\alpha}$$

With $\alpha \approx 2$, this model reproduces the salient features of Bradford's law[12]. The mathematical concepts of growth have become popular through the largely circulated report of the Club of Rome[25]. There are several models which may be applied to discuss / explain the growth of literature. Some of them are discussed below.

THE EXPONENTIAL MODEL

An exponential model[42] is associated with the name of Thomas Robert Malthus[72] (1766-1834) who first realized that any species could potentially increase in numbers according to a geometric series. Exponential growth represents an increase with a fixed proportion of total population for each unit of time. For example, if a species has non-overlapping populations (e.g., annual plants), and each organism produces "b" offspring, then, population numbers "a" in generations t=0,1,2, is equal to:

$$Y_t = a.b^t$$

In this case, the growth rate is (b – 1) * 100. This model is also popularly known as log-linear model, and often expressed as:

$$\ln y = \ln \alpha + \beta x,$$

Here β is the slope of curve, and measures proportional changes in y for a given absolute change in x. The model not only provides the rate of growth (the exponential parameter), but also the rate at which the size of the population doubles. The exponential growth has also been linked to and compared with the size of compound interest. The exponential function assumes a convex shape in its graphical presentation. In exponential growth, the increase is proportional to population size, *i.e.* if the population is y at time t then

$$\frac{1}{y}.\frac{dy}{dx} = \beta$$

where β is the Malthusian parameter of the population. The exponential function can be applied to both growth process as well as decay process. The examples of growth process are given below (Croxton[26])

- Information production process
- Microbiology (growth of bacteria)
- Conservation biology (restoration of disturbed populations)
- Insect rearing (prediction of yield)
- Plant or insect quarantine (population growth of introduced species)
- Fisheries (prediction of fish dynamics). A typical exponential curve is shown below:

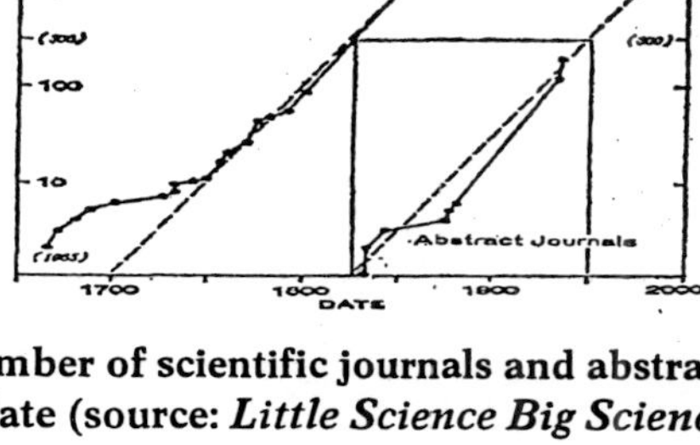

Fig. 5. Total number of scientific journals and abstract journals, as a function of date (source: *Little Science Big Science* by Price[84])

THE LOGISTIC MODEL[68]

The Belgian mathematician Pierre Verhulst (1838) developed the Logistic model. A typical logistic curve is shown in Figure 6. He suggested that the rate of population increase might depend on population density[23]:

$$y = \frac{a}{bc^t}$$

The algebra of the logistic family is something of a hybrid. It mixes together the behaviours of both exponentials and powers (proportions, like rational functions). The three parameters of the logistic family work together to produce its characteristic behaviour, and they are best understood in combination. The parameters b and c are simply the y-intercept and the base of the component exponential function bc^t. The rate at which a logistic function falls from or rises to its limiting value is completely determined by the exponential function in the denominator, by the parameters b and c[68].

Here, as observed in the exponential function, one notices that logistic function depends on the populatior itself, i.e., $\frac{dy}{dx} = \lambda(1 - \frac{y}{k})y$. Pearl and Reed have used the curve to describe the growth of albino rat and tadpole's tail, the number of yeast cells, and most interesting of all, the number of human beings in a geographical area. In 1920, Raymond Pearl and Lowell J. Reed developed it independently. The curve is therefore also known as Pearl and Reed curve. A typical logistic curve is shown in Figure 6. The curve depicts three things:

1. Growth in the early stage.
2. Intermediate period of rapid growth; and

3. An approach to maturity.

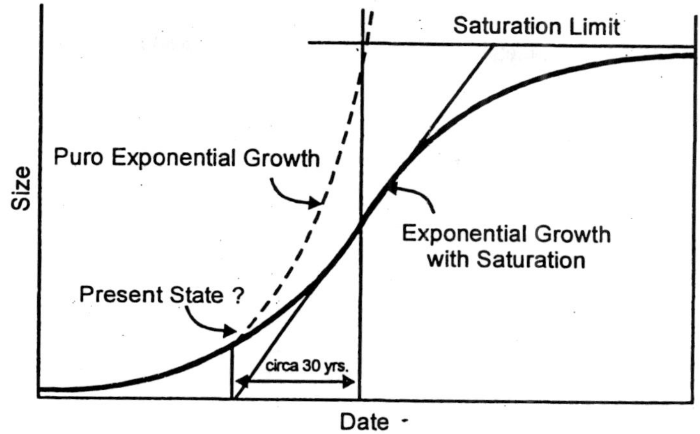

Fig. 6. A typical logistic curve.
(Source: *Little Science Big Science* by Price[84])

THE POWER MODEL

It is a log-log or double log model. It is mathematically represented as:

$$y = \alpha t^{\beta}$$

i.e., $\log y = \log \alpha + \beta \log t$

Sometimes, the power model is also represented as

$$y_t = \alpha + \beta t^{\gamma}$$

where $\alpha, \beta > 0$. For $0 < \gamma < 1$, the function y takes a concave shape, but without an upper limit. For $\gamma = 1$, the function y takes a linear shape. For $\gamma > 1$ the function y takes a convex shape. A typical Power model is shown in Figure 7.

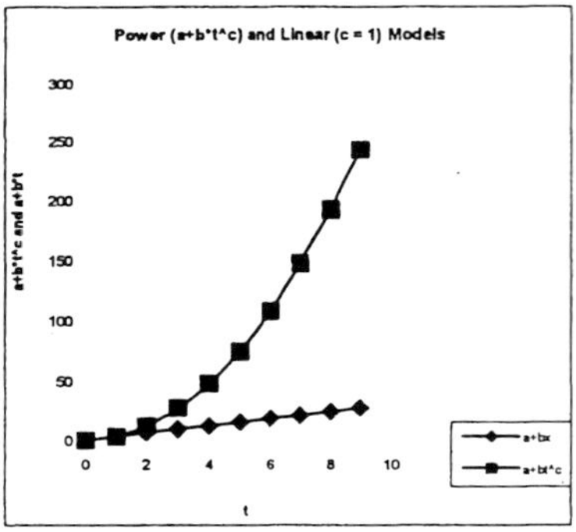

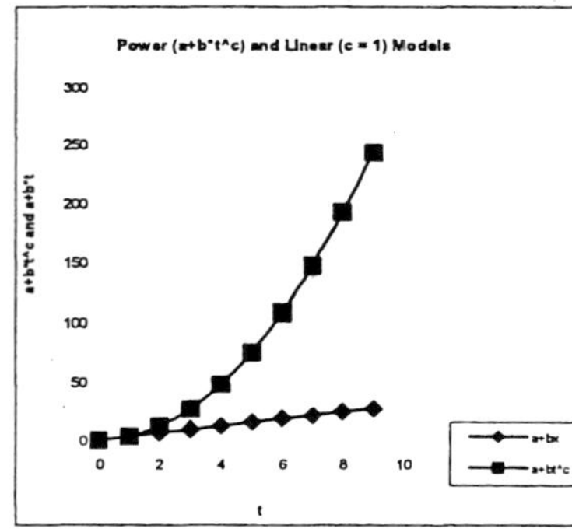

Fig. 7. A Typical Power Model.

THE GOMPERTZ MODEL

The Gompertz model[23,45] describes a trend in which the growth increment of the logarithms is declining by a constant percentage. Thus, the natural value of the trend would show a declining ratio of increment, but the ratio does not decrease by either a constant amount or a constant percentage. The equation for the Gompertz curve is

$$y = ab^{c^t}$$

i.e., $\log y = \log a + (\log b)\, c^t$

The Gompertz and logistic curves are similar in that they both can be used to describe an increasing series, which is increasing by a decreasing percentage of growth, or a decreasing series, which is decreasing, by a decreasing percentage of declines. They differ in that the Gompertz curve involves a constant ratio of successive first differences of the log y values, while the logistic curve entails a constant ratio of successive first differences of 1/y values. The other equally well-known model is Ware's model. It is represented as[35]

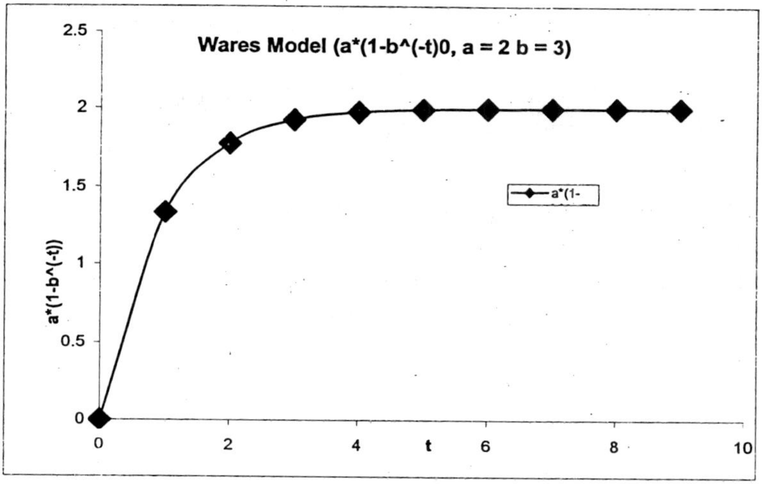

Fig. 8. A Typical Ware's Model.

$$y = \delta(1-\phi^{-t}) \quad \delta, \phi > 1.$$

SELECTING A TREND TYPE

There are many trend types. It is difficult to decide which one to use. Following are some guidelines to select the appropriate curve[23]:

1. If the approximate trend, when plotted on semi-logarithmic paper is a straight line, use an exponential curve.
2. If the first difference resembles a skewed frequency curve, use a Gompertz curve
3. If the first difference resembles a normal frequency curve, use a logistic curve.
4. If the first difference of logarithms are constant, use an exponential curve.
5. If the first differences of logarithms are changing by a constant percentage, use a Gompertz curve.
6. If the first differences of reciprocals are changing by a constant percentage, use logistic curve.

7. It the approximate trend value (or the original data), when expressed, as percentage of a selected asymptote, appears linear on arithmetic probability paper, use a logistic curve.

Egghe and Ravichandra Rao[35] have suggested a useful methodology for identification and classification of growth models. They have classified the growth models based on growth functions, i.e. α_1 and α_2. The growth function is denoted as C(t) (theoretical or concrete data). These growth rate functions may be defined as:

$$\alpha_1(t) = C(t+1)/C(t) \text{ and } \alpha_2(t) = C(2t)/C(t)$$

$$t = 1,2,3, \ldots.$$

α_1 is called the first growth rate function and α_2 the second growth rate function. The basic idea here is that the graphs of α_1 and α_2 are much more different than the corresponding graphs of other growth models. The theoretical relationship between α_1 and α_2 has been worked out to be:

$$\alpha_2(t) = \alpha_1(2t-1)\,\alpha_1(2t-2) \ldots \alpha_1(t)$$

Table 1. Classification of growth models using growth rate function

Types of model	*Growth rate function* α_1	α_2
(1)	*(2)*	*(3)*
Exponential	Type 2	Type 1
Logistic or Gompertz ($0<b, c<1$)	Type 3	Type 4
Gompertz ($b, c > 1$)	Type 1	Type 1
Power ($\alpha > 0, 0 < \gamma \leq 1$)	Type 3	Type 1
Power ($\alpha > 0, \gamma > 1$)	Type 4	Type 1
Power ($\alpha > 0$)	Type 3	Type 2

According to Egghe and Ravichandra Rao, the method of determining the growth goes back to the intrinsic growth rate properties of the data for a good understanding of what is really going on. The authors have also suggested that in order to get a simple clue for the selection of the best model, the plot of two growth rate functions for different mathematical models (namely; exponential, power, linear, logistic and Gompertz) may be drawn and visualized. These graphs can be classified as : Type 1 – increasing, Type 2 – Constant, type 3 – decreasing, Type 4 – increasing and then decreasing, as shown in Table1.

OTHER RELATED WORKS

Crane[22] found linear growth pattern in the growth of literature in two sub-fields, "invariant theory (1887-1941)" and "reading research (1881-1957)". Tague *et.al*[108] explored the cumulative growth of literature as reflected in *Chemical Abstracts* (1907-79); *Science Abstracts* (1960-79) and *Biological Abstracts* (1960-70). They found that the linear growth pattern fits best in the majority of the literature covered. The exponential growth model fits the best in literature covered by the *Chemical Abstracts.* (1907-1979).

May[75] studied the growth of literature on mathematics from 1965 to 1986. He observed that it is an exponential growth, with only few deviations observed during the world wars. Menard[76] has examined the literature in various sub-fields of earth sciences. He found that in the field of vertebrate paleontology, which started as a discipline in the sixteenth century, grew slowly until the end of the eighteenth century, and then began to grow exponentially. Wolfram *et.al*[113] explored the Linear, Exponential and Power models applied to the growth of publications during a period of 20 years, as reflected in the databases of science, technology, social science and humanities. They found that, in most cases, the mathematical model that provided the best fit to the observed data was a power model, rather than an exponential, logistic or a linear model; and they concluded, "The breakdown in exponential growth is well underway. The power model was in particular best, because it has the advantages of modeling the growth behaviour of both the linear and exponential models.

Egghe and Ravichandra Rao[35] clarified the formal distinctions between the four models that Wolfram *et al.* examined, pointing out that any linear model should more properly be recognized as a power model of a special kind, and introducing two other comparable models, the Gompertz and Ware functions, that Wolfram *et al.* did not consider. Revisiting the data collected in the earlier study, Egghe and Ravichandra Rao[35] observed that an exponential model was never the best fit. Indeed, they have shown that such a model could never have been expected to provide the best fit, given that the rate of growth in every database declined steadily over the years studied. They also found that a power model fitted best in cases of convex growth and that a Gompertz model generally fitted best in cases of S-shaped growth. Egghe and Rao's findings suggested that, in modeling the growth of literature, the choice between an exponential and a logistic function may have always been a false one, and that we should instead be asking whether growth is best described by a power law or a Gompertz function. Gupta, and others[50,52] found best fits with either power or logistic models, and called into question the utility of the Gompertz function for modeling S-shaped growth in the social sciences. Gupta, *et.al* have applied this methodology in the application of selected growth models to the growth of world and Indian physics literature during 1898-1950. They observed that growth of Indian physics literature follows a logistic growth model, while the growth of world physics literature can be explained by the combination of logistic and power models

Sharma, *et.al*[105] have examined the growth of world literature as reflected in three data sets, namely, *Physics Abstracts, Chemical Abstracts* and *Electrical and Electronics Abstracts* from 1907-1994. They found that the power model describes best the growth of literature as reflected in the three data sets. Tsay and Yang[109-10] studied the randomized controlled trial (RCT) literature and observed that it plays a fundamental role in developing Evidence-based Medicine (EBM). They found that the literature growth rate, from 1965 to 2001, was steadily rising and followed an exponential model.

Matia *et.al*[73] analyzed a set of three databases at different levels of aggregation: (a) a database of approximately 10^6 publications published from 1980-2001, (b) a database of 508 academic institutions from the European Union (EU) and 408 institutes from the United States for the 11-year period of 1991-2001, and (c) a database of 2,330 Flemish authors published in the period 1980-2000. At all levels of aggregation they found that the mean annual growth rate of publications is independent of the number of publications of the various units involved. They also found that the standard deviation of the distribution of annual growth rates decays with the number of publications as a power law with exponent approx. 0.3. Tsay studied the growth pattern, journal characteristics, and author productivity of the subject indexing literature from 1977 to 2000, based on the subject search of a descriptor field in the *Library and Information Science Abstracts* (LISA) database. The logistic curve fitted well the literature growth from 1977 to 2000 in subject indexing. Gupta and Karisiddappa[51] introduced different approaches for studying the growth of scientific knowledge, as reflected through publications and authors. They applied selected growth models to the cumulative growth of publications and authors in theoretical population genetics from 1907 to 1980. They concluded that among the models studied, the power model is the one which best explains the cumulative growth of publication and author counts in theoretical population genetics.

Jing and Kang[59] analyzed three classical models for the growth of scientific literature: the exponential, logistic and the linear growth model. They also discussed the limits and scope of the logistic model. Ramakrishna and Pangannaya[89] examined the *Derwent Biotechnology Abstracts* and the journal *Animal Cell Biotechnology* with the objectives of studying the relative growth rate (RGR) and doubling time of animal cell culture technology literature between 1983 and 1993. They found a reducing trend of RGR and increasing rate of doubling time. It indicates that the growth is neither exponential nor linear. The size of the literature was calculated by applying the logistic growth formula.

Maheswarappa and Ningoji[71] studied the growth of literature in the field of applied sciences in India based on *Indian Science Abstracts* from 1965 to 1989. They observed that the relative growth rate had declined and it was heading towards saturation. The doubling time of literature has consistently increased. The growth of literature in the field of applied sciences in India does not appear to fit the modified exponential, logistic and linear curves. Parvathamma, Gunjal and Nijagunappa[82] conducted a study in which the growth rate of Indian earth science literature was determined by calculating relative growth rates and doubling time for the period 1978-1988. Results showed that the mean relative growth rate has declined whilst mean doubling time has increased. They suggested that Indian earth science literature follows a logistic pattern of growth.

RELATIONSHIP BETWEEN GROWTH OF SOURCES (JOURNALS) AND ITEMS (ARTICLES)

Archibald and Line[4] with a view to examining the conjecture that scholarly and scientific journal literature is growing exponentially, a study was made of a sample of 190 journals that started before 1950, twenty in each of the nine subject fields and ten in belles lettres The number of articles in each journal between (1950-1980 and 1987) was also counted. The analysis showed a rapid growth in most subjects up to 1970, a much slower growth during 1976-1980, and a slow growth or decline between 1980 and 1987. In general and in physical science and technology, the growth rates declined during 1980 to 1987. Although the total number of journals is increasing, the overall rate of growth of the total number of journal articles is slow. Persson, Glänzel and Danell[83] presented interesting "universal" data (i.e. based on all papers indexed in the volumes 1980-2000 of the *Science Citation Index*® (SCI)). They observed that, between 1980 and 1998, the number of articles increased by (roughly) one third whereas the number of citations received by them has increased by three quarters.

Egghe[28] observed that growth rates of sources (journals) usually are different from growth rates of items (articles); further he argued, "The references in publications grow with a rate that is different (usually higher) from the growth rate of the publications themselves." His study showed that Naranan's model (exponential model: $y=ac^t$) hardly fits the empirical data. He showed that the "simple" 2-dimensional informetrics models of source-item relations are not able to explain this. He has further shown that a linear 3-dimensional informetrics (i.e. adding a new source set) is capable of modeling disproportionate growth. The explanation consists of "defining" a set of "super sources" which produce the original sources but which also attach the items into the original sources. In this way, disproportionate growth of references versus articles can be explained by considering the growth in the number of authors. In the same way, disproportionate growth of articles versus journals (a new dataset is compiled from the database Econlit) can be explained by considering journal publishers. Formulae of such different growth rates are presented using Lotkaian informetrics and new and existing data sets are presented and interpreted in terms of the linear 3-dimensional model. Sahoo[102] in his thesis compared the growth of the journals with that of the articles, in the area of software studies. His data on the number of journals and number of articles published from 1989 to 2003 are given in Table 2. He has computed the growth rates for both journals and articles at the national and world levels. They are given in Table 3. He observed that the correlation coefficient r between the number of journals and the articles is 0.9811. That is, 96% of the variation in y (articles) is due to the variation in x (journals) – higher the number of journals, the higher the number of articles.

However, in the case of world literature he observed that the correlation coefficient between number of journals and the number of articles published is only 0.8517. Unlike India, the correlation is not so high for the world literature. Figures 9 and 10 show the growth rate curves for journals and articles published by the journals for India and world literature respectively from 1990 to 2003. It is observed that the growth rates of journals have decreased with the decrease of growth rate of articles. There is negative growth during 1990, 1992, 1995, 1998, and 2001 for India; for world literature it is during 1991, 1992, and 1998. Some other years like 1995 and 2001 for India, it was observed that when there is a negative growth

Table 2. Distributions of Journals and Articles for India and World.

Year	India				World			
	Journal	C (Journal)	Article	C (Article)	Journal	C (Journal)	Article	C (Article)
t	j	C(j)	a	C(a)	j	C(j)	a	C(a)
(1)	(3)	(4)	(5)	(6)	(7)	(8)	(9)	(10)
1989	164	164	377	377	2283	2283	26280	26280
1990	162	326	298	675	2293	4576	27545	53825
1991	181	507	302	977	1965	6541	23385	77210
1992	138	645	244	1221	1057	7598	17026	94236
1993	134	779	298	1519	1291	8889	16693	110929
1994	209	988	412	1931	1686	10575	23355	134284
1995	208	1196	414	2345	1727	12302	24976	159260
1996	217	1413	410	2755	2210	14512	28203	187463
1997	273	1686	542	3297	2492	17004	41066	228529
1998	236	1922	463	3760	2379	19383	34737	263266
1999	268	2190	565	4325	2391	21774	34813	298079
2000	319	2509	614	4939	2480	24254	43737	341816
2001	316	2825	710	5649	2589	26843	46306	388122
2002	326	3151	631	6280	2724	29567	43132	431254
2003	442	3593	981	7261	2888	32455	67260	498514

in journals, it does not show any negative growth for the articles. From these observations Sahoo concluded that only in some cases the growth of journals affect the growth in the number of articles. The growth rate of journals and of articles are not the same; for the Indian literature average growth rate is 9% for both journals and the articles and for the world literature the average growth rate of journals is 3% and average growth rate of articles is 9%.

Table 3. Growth Rates of Journals and Articles (in %)

Year	*India*		*World*		*Year*	*India*		*World*	
	Journal	*Articles*	*Journals*	*Articles*		*Journal*	*Articles*	*Journals*	*Articles*
1990	-1.22	-20.96	0.44	4.81	1998	-13.55	-14.58	-4.54	-15.41
1991	11.73	1.34	-14.3	-15.1	1999	13.56	22.03	0.5	0.22
1992	-23.76	-19.21	-46.21	-27.19	2000	19.03	8.67	3.72	25.63
1993	-2.9	22.13	22.14	-1.96	2001	-0.94	15.64	4.4	5.87
1994	55.97	38.26	30.6	39.91	2002	3.16	-11.13	5.21	-6.85
1995	-0.48	0.49	2.43	6.94	2003	35.58	55.47	6.02	55.94
1996	4.33	-0.97	27.97	12.92	Avg Growth Rate	9.02	9.24	3.65	9.38
1997	25.81	32.2	12.76	45.61					

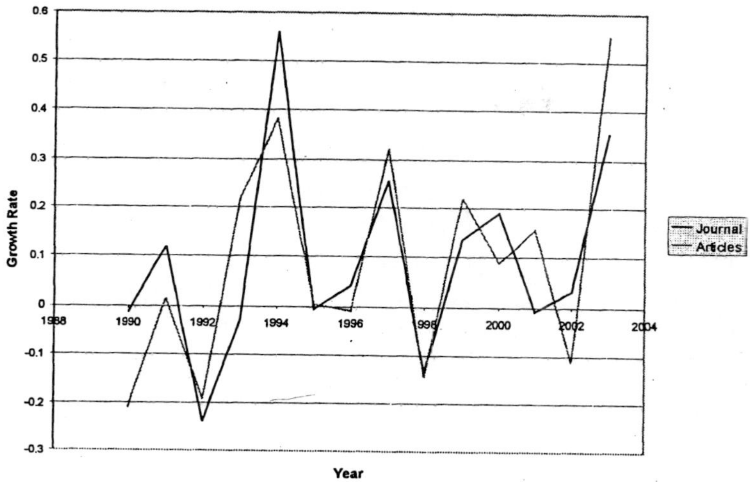

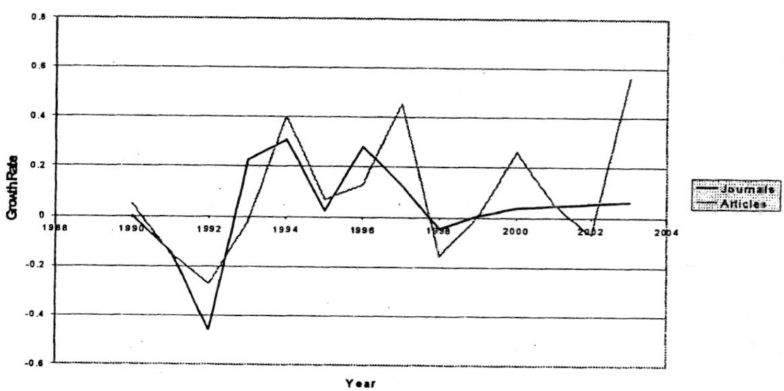

Fig. 9 and 10. Growth rates of Journal Vs Article for India and World literature

Ravichandra Rao and Divya[99] studied the growth of literature in Malaria research. They also studied the relationship among journals, articles and authors. They observed that the number of journals, articles, and authors increased approximately exponentially. The number of articles in malaria research increased from 3,996 to 57, 627 from 1955-1965 to 1996-2005. Also the number of journals has been increased 503 to 3,072 from

1955-65 to 1996-2005. Figures 11, 12 and 13 clearly show that the exponential model fits the empirical data. The R^2 value for the trend for journals, articles, and authors are 0.9502, 0.9475 and 0.9651 respectively; the low R^2 value is perhaps due to the fewer number of data sets; the Figures 11, 12 and 13 are much more convincing that the data on journals, articles, and authors increase exponentially. Under the assumption that the data conform to the exponential model, the growth rates have been computed; the growth rates of the journals, articles and authors are 5.31%, 7.38%, and 10.06% respectively. The most important observation is that the number of least productive journals increased to 2,951 from 463. This was perhaps due to

- Interdisciplinary nature of research in Malaria and related topics
- High growth rates (exponential) of journals and articles

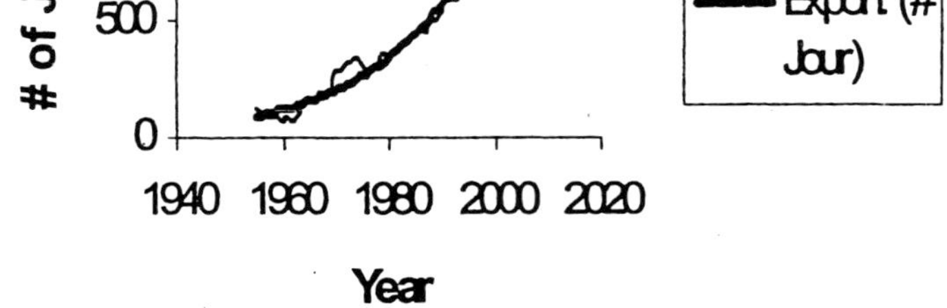

Fig. 11. Growth Curve of Journals
(Source : Ravichandra Rao and Divya[99])

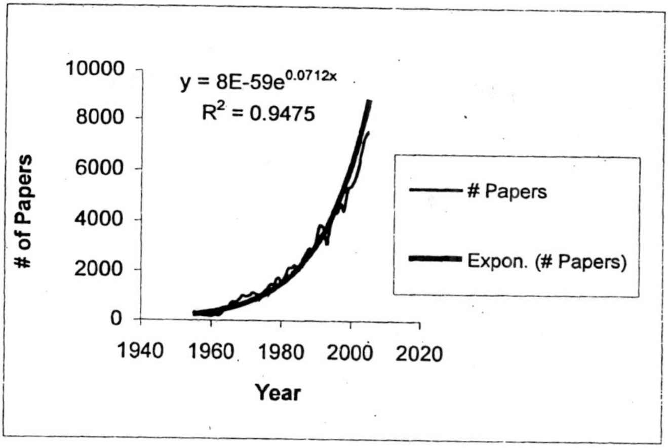

Fig. 12. Growth Curve of Articles
(Source : Ravichandra Rao and Divya[99])

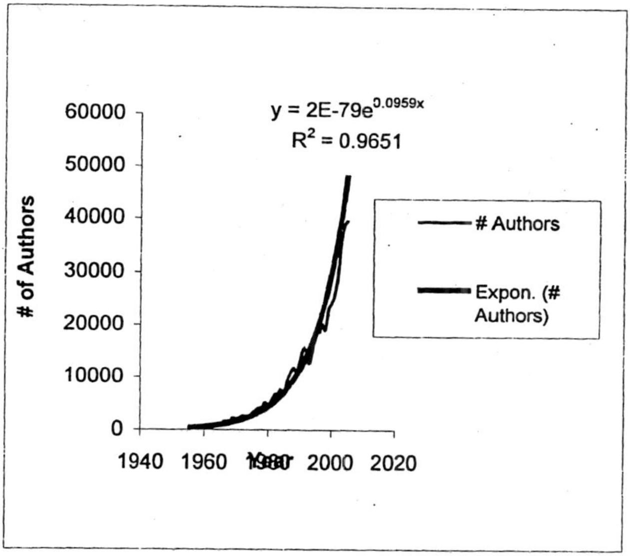

Fig. 13. Distribution of Authors
(Source : Ravichandra Rao and Divya[99])

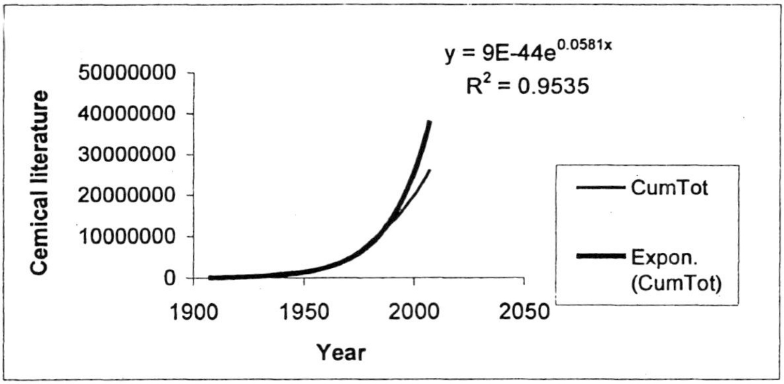

Fig. 14. Growth of Chemical Literature.

An attempt was also made to study[99] the relationship among the three variables – journals, articles and authors. For this purpose a multiple regression analysis was carried out. It was observed that the following multiple regression line fits very well the observed data on journals, articles and authors for a periods of fifty years.

Articles = -39.2771 + 3.61719*journals + 0.085882*Authors

($R^2 = 99.1616\%$)

With regard to this linear equation, the p-value for each of the independent variables (journals and authors) is 0.0000. Since it is less than 0.01, the highest order term is statistically significant at the 99% confidence level. Consequently, there is no need to remove any variables from the model!

An attempt was also made (by me) to analyze the chemical literature based on the data from *Chemical Abstracts.* Figure 14 shows the growth trend with its R^2. The growth rate is 5.98%.

OBSOLESCENCE OF LITERATURE

'Obsolete' generally means out of date or no longer in use. The process of becoming obsolete is known as obsolescence. It is also often referred to as 'phenomenon of replacement.' The term obsolescence was used for the first time by Gross and Gross[47] in 1927. They analyzed the references in the 1926 volume of the *Journal of Chemical Literature* and observed that the number of references fell to one-half in fifteen years. Obsolescence is thus a characteristic of scientific and technical literature. Burton and Kebler[17]were the first to use the term 'half-life' as applied to documents in 1960. It is defined as 'the time during which one-half of all the currently active literature published.' It is the period of time needed to account for one-half all the citations received by a group of publications. The concept of half-life is always discussed in the context of diachronous studies. More precisely, Line and Sandison[66] refer to diachronous studies in those that follow the use of particular items through successive observations at different points in time, whereas synchronous studies are concerned with the plotting the age distribution of material used at one point of time.

However, there is no reason to suppose that the half-life for some subject is the same as the median citation age in that subject. Half-life in the context of synchronous data is referred to as median age of the citations / references. The use of literature may decline much faster with data of ephemeral relevance, if it is in the form of reports, thesis, advance communication or pre-print and in the context of advancing technology. However, the use of literature may decline slowly when it is descriptive (e.g., taxonomic botany) and critical (e.g., literary criticism).

Brookes[13] in one of his articles argued that if growth rates of literature and contributors are equal then the obsolescence rate remains constant. In this sense growth and obsolescence are related. Ravichandra Rao and Meera[94] studied the relationship between growth and obsolescence of literature, particularly in mathematics.

Gupta[49] studied the relationship between growth rates and obsolescence rates and half-life of theoretical population genetics literature. He observed that the lognormal distribution fits very well to the age distribution of citations over a period of time.

In the analysis of obsolescence, Brookes[13] argued that the geometric distribution expresses the idea that when a reference is made to particular periodical of age t years $(1\text{-a})\, a^{t-1}$. 'a (< 1)' is a parameter – the annual aging factor; it is assumed to be constant over all values of t. Let $U = 1 + a^2 + a^3 + a^4 + \ldots. + a^t + \ldots.$ i.e., $U = 1/(1\text{-a})$. Similarly if $U(t) = a^t + a^{t+1} + a^{t+1} + a^{t+2} + \ldots\ldots = a^t\,(U(0)$, then $U(t)/U(0) = a^t$. Using this relation, by graphical method, we can compute half-life as well as 'a'.

If we assume that the literature is growing exponentially at an annual rate of g, we then have $R(T) = R(0)e^{gT}$, where R(T) is the number of references made to the literature during the year T. We also have

$$U(0) = R(0)/(1\text{-}a_0) \text{ and } U(T) = R(T)\,)/(1\text{-}a_T)$$

Where a_0 and a_T are the annual aging factors corresponding to the years 0 and T respectively. Under the assumption that utility remains constant $(U(0) = U(T))$, we have

$$R(0)/(1\text{-}a_0) = R(T)\,)/(1\text{-}a_T)$$

By substituting the value of R(T), we get a relationship value between the growth and the obsolescence:

$$e^{gT} = (1\text{-}a_T)/(1\text{-}\ a_0)$$

However, Egghe and Ravichandra Rao[36] showed that the obsolescence factors (aging factors) 'a' is not a constant, but merely a function of time. They have also shown that the function 'a' has a minimum which is obtained at a time t later than the time at which the maximum of the number of citations is reached.

Egghe[26] developed a model to study the influence of growth on obsolescence. He obtained different results for the synchronous and diachronus studies. He argued that for an increase of growth implies an increase of obsolescence for the synchronous case and for the diachronous case, it is quite the opposite. In order to derive the relationship, he also assumed the exponential models for growth as well as for obsolescence. In another paper, for the diachronous aging distribution and based on a decreasing exponential model, Egghe[27] derived first citation distribution. In his study he assumed the distribution of the total number of citations received conforms to a classical Lotka's function. The first citation distribution is given by

$$\phi(t_1) = \gamma(1 - a^{t_1})^{\alpha-1}$$

where γ is the fraction of papers that eventually get cited; t_1 is the time of the citation, 'a' is the aging rate and a is Lotka's exponent. Egghe and Ravichandra Rao[39] in their study in 2002 observed that the cumulative distribution of the age of the most recent references is the dual variant of the first citation distribution. This model is different from the first citation distribution. In another study, Egghe and Rao[38] have shown the general relation between the first citation distribution and the general citation age distribution; if Lotka's exponent $\alpha = 2$, both these distributions are the same. In the same study, they have argued that the distribution of nth citation is similar to that of the first citation distribution. Egghe, Rao and Rousseau[37] studied the influence of production on utilization function. Assuming an increasing exponential function for production and a decreasing one for aging, these authors have shown that in the synchronous case, the greater the increase in production, the greater the obsolescence; however, for the diachronous case it is quite the opposite. This proof is different from the earlier one derived by Egghe[26].

LECTURE 2

Part 1
AUTHOR PRODUCTIVITY

Scientific productivity is frequently measured in terms of published output, since the data on the number of publications by the authors can easily be collected and also are generally quite reliable. However, the quantitative measures are sometimes inadequate indicators of research outputs. We may have to consider some other variables also, such as

- Type of publications (— general articles, publications in conference proceedings, patents, text books, monographs, etc.); and
- Quality of the sources (— peer-reviewed journals, conference proceedings wherein only peer-reviewed articles are considered, reputation of the publishers, number of citations received – by the same journal / same source or by other journals / other sources, impact factor of the journal, age of the journal or source cited, etc.)

Narin[78] reviewed the early studies of scientific productivity and concluded that "scientific talent is highly concentrated in a limited number of individuals". This type of information may be useful to formulate science policy. Price[85] conjectured that the number of the elites in science is small compared to the total number of scientists. Price[85] also claimed "any population of size N contains an effective elite size $\sqrt{N}$" (Price[85], 1971, p.74) Egghe[30,40] has generalized this law as "N^{α} scientists (elite scientist or sources) contribute 100 α% of the total publications (or items)" Taking the clue that "elites in science is small and further assuming that exchange among scientists in general is a rare event," Griffith and others[46] suggested that informal contacts in science might follow a Poisson distribution. They also attempted to estimate the value of the parameter (based on the data collected by Crane[22]) of the Poisson distribution, that would fit into the empirical data on scientists in the area of rural sociology and finite mathematics. They observe that

1. a majority having a low average rate of contact which follow a Poisson distribution; and
2. elites whose frequency of contacts deviates considerably from what the Poisson distribution would predict which is in fact about 8 times higher than the majority.

Shockley[103] argued that a lognormal distribution describes the pattern of scientific productivity. He argued that in a large homogeneous laboratory "there are great variations in the output of publications between one individual and another" (103, p.280) He further argued that the variation in the productivity among scientists can be explained on the basis of simplified models of the concerned mental processes. One of the characteristics of the model is that a large number of factors are involved, so that small changes in each might result in a very large output. For instance, the number of ideas to which a scientist can impart awareness at one time may control his ability to invent, and his rate of invention might increase very rapidly with the number. He observed that the same individuals create new science at a rate at least fifty percent faster than that of others, if the number of scientific publications is used as a measure of productivity. In the same paper, he also studied the relationship between salary and productivity. He showed that rewards in government laboratories do not keep pace with increasing production. In fact, a scientist has to increase his output 30 to 50 percent for 10 percent raise in his salary. Thus he argued that an efficient operation in many government laboratories is difficult because top scientists are poorly paid in comparison with those who work in industry. Therefore very few of the most productive scientists remain in government laboratories.

PROBABILITY DISTRIBUTIONS OF SCIENTIFIC PRODUCTIVITY

There are many approaches apart from Lotka's law for describing scientific productivity. Williams[114] (1944) examined the geometric model and Fisher's logarithmic series in a study of publication patterns of biologists. Simon[106] in 1955 proposed a Beta function, which he calls Yule distribution. Shockley[103] (1957) proposed a lognormal distribution. Price[86] proposed a Cumulative Advantage distribution, similar to a Beta function. Murphy[77] (1973) observed that the distribution of articles of authors in the humanities conforms to Lotka's law. Schorr[104] (1975) discussed a test of Lotka's law in the field of map librarianship. Coile[20] (1976), however, argued that Lotka's law does not apply to either the field of map librarianship or to the field of humanities. For the purpose of testing the hypothesis, Coile used the Kolmogorov-Smirnov test, which is based on the maximum difference between an empirical and hypothetical cumulative distribution. Coile[21] (1977) in another article proposed a continuous distribution (Weibull) for the data collected by Hersh[54] (1942) and compares his results with those obtained by the use of other models. On the other hand, Bookstein[9] argued that the Lotka's law is invariant under the impact of society on scientific productivity.

LOTKA'S LAW

Lotka[67] in 1926 discussed an analysis of the number of publications listed in *Chemical Abstracts* from 1907 to 1916 with the frequency of publications by particular authors. He excluded the names of corporate authors, and only considered the names of authors whose names begin with *A* and *B* as listed in the index. He also made a similar study in the field of physics. He applied the same process to the name index of *Auerbach's Geschichtstafeln der Physik.*, which covers the entire range of history of physics up to the year 1900. For these two sets of data, he computed the theoretical frequencies of publications of authors using the least square method. For large values of x (where x is the number of papers published by an author), the fluctuation is too high, perhaps, owing to the limited number of persons in the sample. He therefore considered the first 17 points of the data in physics and the first 30 points in chemistry. Based on the data, he proposed the following inverse square law of scientific productivity:

$$y_x = \frac{6}{\pi^2 x^2} \quad x = 1,2,3,\ldots; \; a > 0$$

where y_x is the relative frequency of the authors publishing x number of papers. The value of a is found to be 2 for physicists and 1.89 for chemists. This difference in the value of α is possibly due to the sampling error, especially in the data pertaining to chemistry. Lotka's law is based on a meager set of data and it has not been statistically tested. It is more a conjecture than a law.

If N is the total number of authors, Ny_1, in Lotka's equation, gives the number of authors who have published single paper each. Thus Lotka's equation is determined in its general form by three parameters:

1. The number of scientists with minimal productivity (authors with single paper each - Ny_1);
2. The maximal productivity of scientist (x_{max}); and
3. The characteristic of exponent a

Lotka's equation can also be written in the following form:

$$y_x = k/x^{\alpha}$$

$$x = 1,2,3,\ldots$$

$$\alpha = 2, k = 6/\pi^2$$

Lotka's equation further suggests that the proportion of authors with single papers (k) is a function of α. That is,

$$k(\alpha) = \left[\sum_{x=1}^{x_{max}} \frac{1}{x} \right]^{-1}$$

This suggests that increase of α is accompanied by the increase of low productivity scientists. This implies that for a given N and for a large value of α, the proportion of highly productive scientists will decrease. The higher the value of parameter α, less the number of most productive authors. on patterns of scientific productivity. Ravichandra Rao[90] in his analytical study showed that the negative binomial distribution describes a pattern of scientific productivity under the "success-breeds-success" condition in a wide variety of social circumstances.

METHODS OF COUNTING

A ratio between the output of scientific research and its inputs is a simple measure of scientific productivity.

The other commonly used measures are counting of the publications and citations to publications. The former is a gross indicator of quantity; the latter is a measure of quality of the publications. No doubt, counting of the publications gives rise to simple measures but it also depends upon the method of counting. In particular, three "counting schemes" are used for publications. Egghe[29] and Lindsey[65], mention three methods of counting the number of publications. They are:

1. Method of total counts or normal counting or standard counting – assigning every author a weight one for each of his or her publication during a time period, irrespective of whether he or she is a first author or a second author, etc. One full unit is assigned for author's name, regardless of the number of co-authors present.
2. Method of straight counting – assigning only the first author a weight one for each of his or her publication and for other authors in the same publication a weight zero. This is clearly an imperfect system since the first author only gets a positive credit. Such a methodology may be meaningful, if the authors (in a paper) are listed based on their merit of contribution. If authors are listed according to the alphabetical sequence of their respective names, then this method leads to a wrong set of data with much bias. In deriving the law of Inverse Square, Lotka adopted this method while collecting the data from the Author Index of *Chemical Abstracts* and *Auerbach's Geschichtstafeln der Physik*. The Science Citation Index presents citation data by first authors only (this has been now improved in the Web of Science (Web of Knowledge), where one can search on every author). In another method of counting, namely

"modified straight count", each paper is counted as a unit and only one of the authors (either first or second or third etc.) is given a point instead of giving a point only to the first author of the paper[115]. The assignment of the number is purely based on the judgment of the bibliometrician as to whom he/she has to give one point. This procedure was recommended as an auxiliary one, to be applied in combination with normal counting.

3. Method of fractional counting – In this scheme, every author in a multi-authored paper gets a fraction of the credit, such that all the authors together get a credit of one unit. If n authors have contributed to a paper we then assign every author a weight 1/n. Productivity of the given author is thus expressed as a score obtained by summing only the corresponding parts of each publication.

There are also other methods of counting — they are proportional author count, pure geometric author count, noblesse oblige, etc.

PROPORTIONAL AUTHOR COUNT[30]

It is less used. If an author has rank r in the author list of an article (with A authors in total, hence $r \in \{1, 2, \ldots, A\}$) then this author receives a credit of

$$\frac{2}{A}\left(1-\frac{r}{A+1}\right)$$

It is easy to see that the above formula is the normalized form of a score A +1 – r (for r = 1: the score is A,...,r = A: the score is 1) so that the sum of all numbers in the above equation for r = 1, 2, ..., A, is 1. This method also gives an equal weight of 1 to each article, independent of the total number A of authors.

PURE GEOMETRIC AUTHOR COUNT[41]

If an author has rank r in a list of authors (with A authors in total), then this author receives a credit of

$$\frac{2^{A-r}}{2^A - 1}$$

So this method also gives an equal weight of 1 to each article, independent of the total number authors.

NOBLESSE OBLIGE[30]

As in senior author counting, this author receives a score of $\frac{1}{2}$ while the other A-1 authors receive an equal credit of $\frac{1}{2(A-1)}$. This method also gives rise to total weight of an article to 1, independent of the total number A of authors.

FRACTIONAL COUNTING METHOD

Rousseau[100] in 1992 in his article entitled "Breakdown of the Robustness Property of Lotka's Law: The Case of Adjusted Counts for Multi-authorship Attribution", discussed frequency distribution of "fractional scores" in a bibliography of informetrics. He observed that fractional counting of authors does not lead to a Lotka distribution. He further argued that Bookstein's robustness property of Lotka's law breaks down in such cases. Ravichandra Rao[92] also studied a distribution of fractional scores in mathematics, based on the articles covered in *Math Reviews* (1990.) For the appropriate groups or classes of fractional scores, he observed that lognormal distribution fits much better than other distributions; however, this hypothesis was rejected when appropriate tests were applied. Burrell and Rousseau[16] present numerical simulations of fractional frequency distributions showing their irregular shapes. It is clear that calculation of fractional frequency distribution is very difficult. Egghe and Ravichandra Rao[34] further analyzed this data to present a very good model to describe distribution of fractional scores. Their paper "Duality Revisited: Construction of Fractional Frequency Distributions based on two dual Lotka's Laws," is the first attempt of this kind. They assumed two simple Lotka distributions with exponent 2 – one for the number of authors with n papers (total count) and the other for the number of papers with n authors. Based on the earlier convolution model of Egghe[29], these authors have reworked for discrete scores and produced a theoretical fractional frequency distribution ($\phi(q)$, q refers to fraction of a paper by an author.) with only one parameter. It is in very close agreement with observed data, produced earlier by Ravichandra Rao[92]. The $\phi(q)$ was ultimately computed using $\phi(n)$ (distribution of papers) and $\psi(n)$ (distribution of authors) Further, they also noticed that a Poisson distribution (for distributions of papers), if the parameter λ in Poisson distribution is chosen in the appropriate way, fits fairly well the frequency of fractional scores. They however did not publish this finding..

By denoting $f_1(z)$ as the fraction of authors with fractional score z in 1 paper, Egghe and Ravichandra Rao[34] have shown that

$$\phi_1(z) = \psi(1/z)/\mu z$$

If ψ follows Lotka's law, then for $z = 1$

$f_1(1) = 6/(\pi^2\mu)$; m is the average number of papers per author.

If ψ follows a Poisson distribution, then for $z = 1$

$f_1(1) = e^{-\lambda}/(\lambda+1)$; l is the parameter of the Poisson distribution.

If ψ follows a negative binomial distribution, then for $z = 1$,

$f_1(z) = p^{k+1} / (kq+p)$; p and k are the parameters of the negative binomial distribution.

In order to compute f(q) (number of authors with q papers – q (a real number > 0) refers to a fractional score; i.e., if $q = 0.5$, then the author has 0.5 papers to his credit.), one needs to have the following four functions.

1. Fraction of authors with the fractional score 1 ($f_1(1)$)
2. The author distribution of fractional score in 1 paper ($g_1(.)$)
3. Distribution of papers ($\phi(n)$: number of authors with n papers)
4. Distribution of authors ($\psi(n)$: number of papers with n authors)

With these four functions, Egghe and Ravichandra Rao[34] have derived a theoretical model for the fractional frequency distribution f(q) (discrete case) from two dual Lotka's laws. They have derived the required formula for f(q), $q > 0$, for different cases:

1) Case 1: i = 2, allowing an author score of 1/2 or 1 in one paper
2) Case 2: i =3, allowing an author score of 1/3,1/2 or 1 in one paper
3) Case 3: i =4, allowing an author score of 1/4, 1/3,1/2 or 1 in one paper
4) Case 4: i = 5, allowing an author score of 1/5 1/4, 1/3,1/2 or 1 in one paper

For each case i, fractional scores of 1/j for j larger than or greater than i are set to 1/i for reasons of simplicity.

The case 1 for i = 2 is the simplest one and the relevant formulae are :

$$f(1/2) = g_1(1/2)\,\phi\,(1) = (1\text{-}f_1(1)\text{-})\,\phi\,(1)$$

$$f(1) = g_1(1)\,\phi\,(1) + (g_1(1/2))^2\,\phi(2) = \phi_1(1)\,\phi(1) + (1-\phi_1(1))^2\,\phi(2)$$

$$f(3/2) = 2g_1(1/2)\,g_1(1)\,\phi(2) + (g_1(1/2))^3\,\phi(3)\text{-}$$
$$= 2(1\phi_1(1))\phi_1(1)\,\phi(2) + (1-f_1(1))^3\,\phi(3)$$

$$f(2) = (g_1(1))^2\,\phi(2) + 3(\,g_1(1/2))^2\,g_1(1)\,\phi(3) + (\,g_1(1/2))^4\,\phi(4)$$
$$= (f_1(1))^2\,\phi(2) + 3(1-f_1(1))^2\,\phi_1(1)\,\phi(3) + (1\text{-}f_1(1))^4\,\phi(4)$$

Formulae for case 2 (for $i = 3$) and for case 3 (for $i = 4$) are given by Egghe and Ravichandra Rao[34]. They have also computed the formulae for case 4 ($i = 5$) (not published)

DERIVATION OF $g_1(.)$

$g_1(.)$ the author distribution of fractional scores in 1 paper.

A: Computation of $g_1(.)$ where $\psi(n)$ confirms to Lotka's law

Case 1: i=2

This is a simple case. An author receives a score 1, if he / she is an author in a single authored paper. If he/she is an author in a multi-authored paper, the author receives a score 1/2.

$$g_1(1) = f_1(1) = 6/\pi^2\mu$$
$$g_1(1/2) = 1 - f_1(1) = 1 - 6/\pi^2\mu$$

For formulae for case 2, 3 and 4 for i = 3,4, and 5 respectively refer to (34, 102.)

B: Computation of g(.) where $\psi(n)$ is Poisson distribution

Case 1: i=2

This is a simple case. An author receives a score 1, if

he / she is an author in a single authored paper. If he/she is an author in an multi-authored paper, the author receives a score 1/2.

$$g_1(1) = f_1(1) = e^{-\lambda} / (\lambda + 1) \text{ and}$$

$$g_1(1/2) = 1\text{-}f_1(1)$$

For formulae for case 2, 3 and 4 for i = 3, 4, and 5 respectively refer to (34, 102.)

C. Computation of g(.) where ψ(n) is negative binomial distribution

Case 1: i=2

An attempt was also made to fit both the ψ(n) and ϕ(n) confirms to Negative Binomial distribution. The following formula has been derived and fitted the distribution as follows.

$$g_1(1) = f_1(1) = \frac{p^{k+1}}{kq + p} \text{ and}$$

$$g_1(1/2) = 1\text{-}f_1(1)$$

For formulae for case 2, 3 and 4 for i = 3,4, and 5 respectively refer to (102)

f(q) for different values of q were computed for the following four combinations of ψ(n) and ϕ(n) for cases i to iv. The results are given in (102).

i. Case of both ψ(n) and ϕ(n) follow Lotka's Law.
ii. Case of ψ(n) follows Poisson distribution and ϕ(n) follows Lotka's distribution.
iii. Case of ψ(n) follows Poisson distribution and ϕ(n) follows negative binomial distribution.
iv. Case of ψ(n) follows negative binomial distribution and ϕ(n) follows negative binomial distribution.

Egghe and Ravichandra Rao computed ϕ(q) under

the assumptions (i) and also for (ii); the results were excellent; based on the results for (i), they argued that modeling fractional counting does not prove a breakdown of Lotka's law (as argued by Rousseau[100]; but, on the contrary, Lotka's law $\phi(n)$ and its dual y(n) explain the fractional frequency distribution. The results for (ii) were not published. Rao, Sahoo and Egghe[95] have computed $\phi(q)$ for the software data under the assumptions (i), (ii) and (iii). They observed that the results were good for (iii). Sahoo in his thesis computed $\phi(q)$ for all the assumptions; In his study, he has observed that

- Poisson distribution can be suitable model to explain the distribution of authors ($\psi(n)$)
- Negative binomial distribution is the next best model to explain the distribution of authors ($\psi(n)$)
- Lotka's law certainly does not fit well for the distributions of authors ($\psi(n)$)
- The distribution of papers follows a negative binomial distribution ($\phi(n)$)

Part 2
Distributions of Multiple Authors

Price[84] in 1963 asserted, "since the turn of century the proportion of the multi-authored papers has accelerated steadily and powerfully and it is now so large that if it continues at present rate, by 1980 the single paper will be extinct". In 1910 the proportion of single authored papers stood 95% and since then the overall decline has been fairly gradual in astronomy as compared to chemistry.

Fields% of papers having single authors in the year.

	1910	*1934*	*1940*	*1960*	*1963*
Astronomy	95	89	86	73	68
Chemistry	82	72	67	40	30

Source: JG O'Connor, 1969, *DRTC Seminar* 7, p.463

Cherno Gorenko and Muchnik[19] studied the subject phosphides and observed the increasing trend of numbers of two and three authored papers, between 1961 and 1973. The number of single author papers has decreased to 21 % in 1973 from 70% in 1961. Two and three authored papers increased from 15% to 30% and 37% respectively.

Beaver[7] in his paper "Does collaborative research have greater epistemic authority?" pointed out that collaborative research processes have greater epistemic authority than research produced by single researcher. In his study, he argued that single authored papers are slightly more likely never to be cited than collaborative ones. His data suggests that collaborative research produces significantly more authoritative research." Kundra[62] in his thesis observed an increasing trend towards collaboration and large disparities in the rate of collaboration between 'pure' and 'applied' areas. He investigated collaboration influences in basic and applied fields of medical science in India. His findings clearly showed a higher degree of collaboration, i.e., an average of 21.95 papers in basic fields of medical science in contrast to an average of 7.63 papers per year in applied medicine. It is in contrast to a general belief that applied medicine is more collaborative in nature than basic research in medicine. This however was in line with the findings of scholars who also reported a higher degree of international collaboration in basic sciences than in the applied fields. Lal and Ray[63] have analyzed the data from 511 papers on horticulture subjects published in four leading journals during 1987-88.. The analysis revealed that multi-author papers are more than those of single authors. Single authors' contribution was 17.42%, whereas double authors contributed 35.02%. It shows that the concept of solo research has been replaced by team research, particularly in applied sciences like horticulture.

Murphy[77] has analyzed the data from the journal *Technology and Culture*, a quarterly specializing in the history of technology, for a period from 1960 -1969. He considered only research articles and excluded reviews, short communications, research notes, bibliographies and other occasional features. In this study, there were 130 papers by single authors, 33 by two authors, 24 by three authors, 8 by four authors and 5 by five authors. The tabulation essentially verifies that Lotka's law applies to this field (within the humanities.)

Mendez[74] and others presented an analysis of Spanish bibliography of physics and engineering which was retrieved from the database INSPEC and COMPENDEX covering the period 1978-1983. A total of 4,367 documents corresponding to physics and 1,914 corresponding to engineering were analyzed. It was shown that literature published by universities follows an exponential curve, while that produced by research and other organizations, including industries follows a linear growth. The scientific productivity of industry and research organizations is decreasing in the engineering field. They have examined three aspects of co- authorship index:

1) the habits of co-authorship related to the institutions the authors belonged to,
2) the changes in the co-authorship index with time, and
3) the difference in the index when publishing in foreign or Spanish journals.

Research carried out in the more traditional research centers had a higher co-authorship index. It was also noted that it was higher in foreign than in Spanish papers and was growing with time. All these observations seem to point to a trend towards a better quality. Guan and Ma[48] studied the research performance in the computer science for four major western countries, and India and China on the basis of the INSPEC database during the period of 1993-2002. They observed that scientists in the western countries preferred to work separately or only with one collaborator. Asian scientists preferred "multi-authors" (— three and four authors) and "mega-authors" (— five or more authors) collaboration patterns. Suresh Kumar and Garg[107] analyzed 2,058 papers published by Chinese authors and 2,678 papers of Indian authors in the field of computer science during 1971-2000; India's output is significantly higher than the Chinese output. The proportion of a single author papers in China is greater than India's. The proportion of two-author papers, multi-author, and mega-author papers is greater for India as compared to the data for China. This implies that team research in India is more prevalent than in China.

Vijay[112] highlights the collaborative research and authorship trends in food science and technology in India. He examined and analyzed the authorship patterns in food science and technology periodicals, to (i) study the proportion of single-authored articles compared with multi-authored ones, (ii) determine the degree of collaboration in this field, and (iii) study the publication pattern of authors from different organizations. The results indicated that collaborative research was preferred to solo research in this subject in India and the degree of collaboration was found to be 0.91. The average number of authors/paper also showed an upward trend from 4.89 in 1994 to 8.2 in 2003.

Hartley[53] assessed the refereeing time taken from submission to acceptance for psychology journal articles. Based on the data, he observed that the (i) papers of single authors were refereed faster than were those of pairs or larger groups, and (ii) papers of single authorship that contained portions previously delivered at conferences were refereed fastest of all. Thus, there is some impact of "multiple authorship" even on refereeing / reviewing papers. Bhattacharya and Meyer[8] argued that co-invention analysis may have the basic concept with co-authorship analysis but co-invention data exhibits different characteristics. They have observed that

- the level of collaboration in technology has not reached the level in science; and
- co-invention networks do not resemble co-authorship networks, they are more dyadic in nature and thereby characterized by a few connections only.

Kretschmer *et al*[61] examined the similarities between scientists and Web indicators of collaboration, if any, for possible use in technology, and policy making. She examined Web hyperlinks and Web visibility indicators, to establish their usefulness, indicators of collaboration and to explore whether similarities exist between Web-based structures and bibliographic structures. Her study indicates that hyperlink structure on the Web does not reflect collaborative structures as it appears in bibliographic data. She argued that Web visibility indicators of collaboration are different from hyperlinks and can be successfully used as Web indicators of collaboration. Kim[60] observed, "The proportion of internationally collaborated papers did not increase substantially during the last two decades, which is against the results of other studies". Based on the analysis of the number of participating, countries, Kim showed that, "multilateral collaboration has increased considerably in the last 20 years, though the proportion of international research collaboration remains stable." *Lundberg et al*[69] studied university-industry collaboration, based on co-authorship data. Their results show that one-third of the companies (out of 436 companies) that had provided funding to the universities had not co-authored any publications with the university. Further the funding indicator identified only 16% of the companies that had co-authored publications. They also observed a case of conflicting trends between funding and co-authorship indicators.

Ivanisevic and Sapunar[58] analyzed the number of co-authors in articles published by authors affiliated with domestic (Croatian) and foreign (non-Croatian) institutions in the *Croatian Medical Journal*. Their study was based on 761 articles published in *Croatian Medical Journal* between 1992 and 2003. Li-chun and others[64] argued that co-authorship among scientists represents a prototype of a social network. By mapping the graph containing all relevant publications of members in an international collaboration network (COLLNET), the authors inferred the structural mechanisms that govern the topology of the social system. Further they argued that the structure of the network affects the

1. information available to individuals as their opportunities to collaborate, and
2. overall flow of information, and the nature of the scientific community.

They discussed various measures for both the macro- (whole-network) and micro- (actor-centered) structure of collaboration, and then applied to COLLNET. They observed that the COLLNET community displays many aspects of a "small-world," and is somewhat vulnerable to disruption should major figures become inactive. Ravichandra Rao and Sahoo[96,97] studied:

1) Trend in multiple authorship – in two journals (*JASIST, Scientometrics*) and in the subject "software studies"
2) Inter-country and inter-institutions collaboration.
3) Trend of citations to the same journals, to other journals and to other types of documents.
4) Distribution of multiple authors in the field of software studies.

They collected the data on multiple authors from *JASIST* and *Scientometrics* for 12 years, from 1992 to 2004. The average number of authors per paper in *JASIST* varied from 1.45 in 1992 to 2.29 in 2004. The average number of authors per paper in *Scientometrics* varied from 1.43 in 1992 to 2.35 in 2004 — i.e. collaboration between the authors, may be between the institutions or countries is increasing. For the 12 years period, the average of the average number of authors per paper is $1.94(\bar{x}_1)$ and $1.88(\bar{x}_2)$ for papers in *JASIST* and *Scientometrics* respectively. Further t-test (t-value is 0.6426; t_α (for $\alpha = 0.25$) = 2.07) indicate that there is no difference between the means of the two populations - - in *JASIST* and *Scientometrics.* Kolmogorov Smirnov (D_{max} : 0.0079 where D_α (for a = 0.05) = 0.2776) test further suggests that both the data come from a same "population;" i.e., the distribution of multiple authors in *JASIST* and *Scientometrics* are likely to be same !

The number of papers with authors from different countries (in JASIST) declined from 46.15% in 1992 to 39.71% in 2004; i.e. publications based on collaboration between two countries in these two journals are decreasing considerably. In fact, in *Scientometrics,* it has declined from 39.22% to 21.20 in 2002 (Table-4) In 2004, it was only 28.33. It was also noted that the percentage of authors from different institutions have increased from

a) 48.48% in 1992 to 52.94 % in 2004; in fact it was 56.16 % in 2002 and 56.36 in 1996 – the trend is increasing – in *JASIST*

b) 60.61% in 1992 to 76.47 % in 2004 in *Scientometrics*

Further, it was noted

a) Citations to other journals by articles in *JASIST* varied from 32.58 % in 1992 to 30.32% in 2004; in fact the trend is close to "declining". A similar trend is also noted in *Scientometrics.*

b) Citations to other documents are centered around (on an average) 45% to 50% in *JASIST.* In *Scientometrics,* the figures are 56.53% in 1992 and 37.9% in 2004.

c) Citations to articles in *JASIST* by *JASIST* vary from 5.99% in 1992 to 9.63% in 2004, where as, in scientometrics, it varies from 2.42% in 1992 to 28.68% in 2004.

ANALYSIS OF MULTIPLE AUTHORSHIP (IN SOFTWARE STUDIES WITH REFERENCE TO INDIA)

Sahoo[102] in his study observed that the percentage of single authorship paper is decreasing – from 16% in 1989 to 12% in 2003, with one or two variations in the years, 1990, 1994 and 1999. The following trends were observed from the data set during the years 1989 to 2003:

1) The percentage of single-author papers decreased from 16 % (in 1989) to 12 % (in 2003).
2) From 1989 to 1996 the percentage of 2-authored papers increased; however since 1996, the percentage of 2-author papers has decreased. This is perhaps due to the increase in collaborative nature of research in the field where more than two authors are contributing to the research work.
3) 3-author papers increased from 25% (1998) to 29% (2003) which shows a 4% growth.
4) The percentages of 4-author papers increased it was 8% in 1989 and finally it is about 10% in 2003.
5) 5-author papers have also increased significantly during this period 2.9% (1989) to 5.1% (2003).

At the global level he observed:

1. The percentage of single-author papers have decreased from 38% (1989) to 17% (2003).
2. 2-authored papers do not show a constant growth - during 1989 it was 33% and the percentage increased since 1989 to 1994 up to 36% and afterwards it started decreasing. Finally in 2003 it was about 29%.
3. 3-authored papers increased from 17% (1998) to 25% (2003) with an 8% growth
4. The percentage of 4-authored papers increased from 7% in 1989 to 14% in 2003.
5. 5-authored papers increased significantly, from 2.4% (1989) to 6.5% (2003).
6. The papers having more than five authors also increased significantly.

In order to identify a probability distribution function, Sahoo attempted to fit several probability distributions (—Poisson, Geometric, Lotka, and Negative Binomial) to the empirical distributions of both India data and world data. He argued further

(i) Poisson distribution can be a suitable model to explain the distribution of multiple authorship;

(ii) Negative binomial is the next best model to explain the distribution of multiple authorship; and

(iii) Lotka's model certainly does not fit well –it is not even a good approximation.

Lecture 3

Part 1
Hirsch-index (h-index) And g-index

Hirsch[55] in his paper "An index to quantify an individual's scientific research output" proposed an index called h-index. It is defined as the number of papers (h) with citation number higher or equal to h. i.e., "a scientist has index h, if h of his/her N papers have at least h citations each and the other (N-h) papers have no more than h citations." He has further argued that h is preferable to other single parameter criteria commonly used to evaluate scientific output of a researcher, such as, total number of papers, total number of citations, citations per paper, number of significant papers, number of citations to each of q most cited papers, etc. Each of these parameters has its merits and demerits. The h-index does measure the broad impact of individual's work. He has also further argued "for a given individual, h increases linearly with time".

Many have subsequently worked in this area. For instance, Braun, Glanzel and Schubert[15] applied this concept of computing h-index for journals. They compared the ranks of journals based on impact factors with that based on h-index. They suggested that an h-index would advantageously supplement journal impact factors at least in two aspects – it is robust and it combines the effect of quality and quantity. In their study Bornmann and Daniel[10] observed that "on an average the h-index for successful applications for post-doctoral research fellowships was consistently higher than for non successful applicants". Batista and others[6] argued that "although h-index is robust in several ways but yet strongly dependent on the research field." They then proposed a complementary index $h_1 = h^2 / N_a$, where N_a is the total number of authors in the considered h papers. van Raan[111]. in his study titled "Comparison of the h-index with standard bibliometric indicators and with peer judgment for 147 chemistry research groups" presented characteristics of the statistical correlation between the h-index and several standard bibliometric indicators (SBI) as well as results of peer review judgment. He observed that both h-index and SBI relate in a comparable way with peer judgment. However, he also found that for smaller groups with fewer citations, the SBI appeared to be a more appropriate measure of research performance.

Egghe and Rousseau[32] in their paper "An informetric model for the h-index", extended the definition of h-index to the general framework of Information Production Processes", using a source-item terminology. They showed that in each practical situation an IPP always has a unique h-index (h). In the context of Lotkain model, they have also highlighted the relation between h and the total number of items ($h = T^{1/\alpha}$); if $\alpha =$ 2, it coincides with Price's square root law! Egghe[31] in another paper on 'Theory and practice of the g-index", introduced an index called g-index (g) to measure the global citation performance of a set of articles. It is defined as the largest number such that the top g articles receive at least g^2 citations. He has proved the unique existence of g for any set of articles and g >= h. By definition, g is less than or equal to T_p (total number of papers.) and it can be greater than T_p, in which case (g-T_p) papers are with zero citations each. The index based on this principle is referred to as g'-index in (97). For instance, if $T_c = 2500$, and if T_p, then g = g' is 50; that means there are 28 papers with zero citations each and hence the total number of papers (T_p) becomes 50. It thus inflates the actual value of T_p. Rousseau[101] in his article on "Simple models and the corresponding h- and g-index" discussed the relation between h and g in the context of simple models. Egghe[33] in his paper on "Distributions of the h-index and the g-index" discussed the problem of determining the distribution of these indexes. He has applied linear three-dimensional Lotkaian informetrics to determine the distributions. Under the following two assumptions:

1. Article-citation Information Production Process (IPP) to be Lotkaian with exponent α, and
2. Scientist-article IPP to be Lotkaian with exponent α^*

Egghe has shown that both h-index IPP and g-index IPP are Lotkaian with exponent $\alpha\alpha^*$, for both discrete and continuous cases. He further pointed out that both the distributions are highly skewed, due to the high value of áá*!

Ravichandra Rao[98] studied 'two hundred and fifty two references' from a bibliography on "optical flow estimation[1]" In that bibliography 168 authors contributed to 252 references. For each of these authors, a list of their publications was compiled using Web of Science; for each of these papers, the number of citations received was obtained again using the Web of Science. Based on the following data:

1. The number of papers published (T_p) by each of the 168 authors,

2. The number of citations received by each of the papers for every author (in a set of 168 authors), and
3. Total number of citations (T_c) received by each of the authors, h and g were computed as defined by Hirsch[55] and Egghe[31]. Frequency distributions of h and g are given in Tables 5 and 6.

Data in Tables 5 and 6 suggest that the empirical distributions of h and g are highly skewed. Egghe[33] also argued that the distributions of h and g are highly skewed, due to high values of $\alpha\alpha^{*}$, where α and α^{*} are the Lotkaian exponents in the distributions of articles *vs.* citations and scientists *vs.* articles respectively. The averages ('x) of the distributions of h and g are 3,67 and 5.82 respectively. The standard deviation(s) of h and g are 3.45 and 6.17 respectively. For both the distributions, 'x < s^2; in such circumstances, negative binomial distribution will fit the empirical data fairly well. The values of mean and standard deviations indicate that the empirical distributions of h and g are likely to follow an exponential distribution, since they are approximately equal to each other. We therefore attempted to fit exponential, negative binomial and also lognormal distributions to the observed set of data.

- The exponential distribution fits the observed data for both the distributions of h and g very well.
- The negative binomial distribution fits equally well for the empirical distributions of g. The negative binomial distribution fits the empirical distribution of h only at 0.01 levels.

It is observed that D_{max} is least for exponential distribution as compared to the lognormal and negative binomial distributions. The D_{max} values are 0.0954 and 0.0813, for distributions of h and g respectively. The critical value at 0.05 levels is 0.1049. Therefore,

Ravichandra Rao[98] concluded that the **exponential distribution is a suitable model to explain the distributions of h and g.** However, theoretical explanations are yet to be explored in this regard. An attempt was also made to identify a relation between (T_p & h), (T_p & g),), (T_c & h), (T_c & g). For this purpose, correlation coefficients (r) were computed.

- The values of r for (T_p & h) and (T_p & g), are 0,81756 and 0.801 respectively. These values indicate that there exists only a weak linear relation between the total number of papers and h- or g- index.
- The values of r for (T_c & g) is 0.937, indicating that there exists a weak linear relation between the total number of citations and g.
- The value of r between T_c & h (r = 0.7461) and also between log (T_c) and log (h) (r = 0.8186) are very low.

In order to study whether or not there exists a correlation between h and g, the correlation coefficient was computed. The value 0.9678. (R^2 is 0.9366; i.e., 93% of the variation in g is due to the variation in h.) It indicates that there exists a linear relation between h and g. The correlation between (T_p & h) is very poor – only 66% of the variation in h is due to the variation in p. It clearly indicates that **h is unlikely to increase linearly with p.** Fitting of a non-linear model between h and T_p ($h = T_p^{1/\alpha}$) gives a much better result than fitting \a linear model. (R^2 is 0.7551.) Similarly, fitting a non-linear model between h and T_c ($h = k^*T_c^{1/á}$) gives a much better result than fitting a linear model (R^2 is 0.6822.)

Table 5. The empirical distribution of h

h	*f(h)*	*h*	*f(h)*	*h*	*f(h)*	*h*	*f(h)*
0	6	4	23	8	6	13	1
1	39	5	7	9	2	14	1
2	37	6	7	10	4	18	1
3	22	7	8	12	3	25	1
Total		168		Variance		11.9365	
Mean		3.6667		St. Dev.		3.4549	

Table 6.The empirical distribution of g

g	*f(g)*	*g*	*f(g)*	*g*	*f(g)*	*g*	*f(g)*
0	6	6	9	12	2	20	1
1	21	7	7	13	4	21	1
2	35	8	7	14	2	22	1
3	19	9	6	15	3	25	1
4	13	10	4	16	1	35	2
5	15	11	4	17	3	45	1
Total		168		Variance		38.04928	
Mean		5.82143		St. Dev.		6.1684	
Mean		8.7976		St. Dev.		8.692881	

Part 2
WEBOMETRICS

Webometrics basically deals with quantitative analysis of various characteristics of web resources. Several new terms have been proposed for this topic. For instance, Bossy[11] coined the term netometrics, Abraham[2] used the term webometry, Almind and Ingwersen[3] suggested internetometrics and they again renamed it as webometrics, Chakrabarthy and others[18] suggested web bibliometry, Dhyani and others[24] used the term web metrics. Isidro Aquillo started a journal called *Cybermetrics.* Both webometrics and cybermetrics are now popular and they are used as synonymous. However, webometrics concentrates on quantitative studies of web resources and cybermetrics concentrates on quantitative studies of internet applications.

The science of webometrics attempts to measure the World Wide Web (WWW) to know about the number and types of hyperlinks, structure of the WWW and usage patterns. It is the study of "quantitative aspects of the construction and use of information resources, structures and technologies on the Web drawing on bibliometric and information approaches."

WEB IMPACT FACTOR

The concept of web impact factor (WIF) was first introduced by Ingwersen[57]. It is measured as the number of web pages in a web site receiving links from other web sites, divided by the number of web pages published in the site that are accessible to the crawler. It works well only within a single country's webosphere, using a single subject field. (Noruzi[86]). While computing the WIF, one can even limit the links to inlinks only (no self link).

Webometrics in general aims at designing and developing methodologies to measure visibility, such as, WIF. The WIF provides a way to evaluate a web site's relative importance, especially when we compare it to others in the same field or a country's domains (no comparison in different fields). It is computable in relation to a national, sector, and larger web segments or top-level domains. It provides a quantitative indicator of web site's long-term influence; it simply reflects the ability of web sites and webmasters to attract users. It may provide novel insights into the retrieval process on the web.

Web page inlink rates determine the web site, but not vice versa – the impact factor of web sites is not statistically representative of its web pages. The WIF is a function of the number of inlinks per web site not per web page and also there are international language biases. The original intention for the use of the WIF was to allow comparison between the link rates of web sites. The misuse of this calculation has in recent years widened to include evaluation of the quality of world university web sites ranking. Studies related to ranking of world universities (– ranked according to indicators measuring web presence and impact) are good examples of use of WI. It is not meaningful; it has to be corroborated by other measures.

The WIF can be influenced and biased (intentionally or otherwise) by many factors and variables. It depends on the search engines' web coverage. For example, consider two web sites A & B, wherein, the web site A is linked 10 times and published 100 web resources over the period and web site B is linked 100 times and published 100 web resources – both web sites yield same impact factor. Does it mean that they have same influence? Large web sites that publish many web resources may not have as high an impact as smaller scientific and research web sites, because the high web resources rate counteracts the high inlinks rate (Noruzi)

Relationships between fields strongly determine the WIF — interdisciplinary web sites attract more inlinks; web sites related to research or developing technology usually have larger impact factors. Unlike scientific citations to journals, institutions or individuals, which may be stable or may constantly increase the number of pages linking up to a particular web resource may indeed decrease or disappear over time.

Evaluation of web sites is a difficult task. In this regard, the WIF is not a perfect tool to measure the quality of web sites but there is nothing better at present and WIF has the advantage of already being used. It is a simple and computable index for quantitative evaluation of web sites.

WEB CITATION ANALYSIS

A number of Webometric investigations have focused not only on web sites but also on academic publications; for example, there are instances of using the web to count how often journal articles are cited. The rationale behind this is partly to give a second opinion for the traditional ISI data and partly to see if the web can produce evidence of wider use of research including informal scholarly communication and for commercial applications.

A significant amount of webometrics research has also evaluated commercial search engines. The two main topics have been the extent of coverage of the web and the accuracy of the reported results. However, research related to developing search engine algorithms (IR) and how search engines (SEs) are used (information seeking) are not part of webometrics. It is logical to assess how much of the web they cover? In a survey conducted in 1999, it was observed that none of the SEs covered more than 17.5%. Overlap among the SEs is surprisingly low. The issue of currency of SEs is however multi-faceted. A new study of coverage of the indexable web search engines states that as of Feb 1999 only 42% of the web is indexed by the combined SEs. However, their new study indicates that it has increased to 60%.

MEASURING WEB 2.0

In web 2.0, web sites are mostly driven by consumer content – blogs, Wikipedia and social network sites, particularly in the context of data mining. It should be possible to extract patterns such as consumer reactions to products or world events. To address these issues software has been developed by IBM (Web Fountain), Microsoft (Pulse) etc. A good example of a research initiative to harness consumer generated media (CGM) is an attempt to predict sales patterns for books based upon the volume of BLOG discussions of them (as compared to reviews)

Measuring Web 2.0 – it is unlikely to replace traditional bibliometrics, but it can be useful for several other purposes. It cab be used for pilot studies to identify areas for follow-up systematic bibliometric analyses; to assess the extent to which researchers are successful in publishing their work online; and for relational analyses of communication in disciplinary or geographical areas of science. Its methods can help the analysis of Web 2.0 and online repositories for social sciences and humanities research goals.

Part 3
ISSUES IN SCIENTOMETRICS

Scientometric research includes studies pertaining to scattering of articles over journals, growth of literature, obsolescence of documents, circulation studies, author productivity, impact of research, distribution of publications by country, by language, by institutions, disciplines, type of documents, etc. There are very few articles in bibliometrics concerned with theoretical and philosophical foundations. Some of the important studies on theoretical and philosophical foundations are in the area of:

- Law of Scattering (Bradford's law)
- Author productivity (Lotka's law)
- Word productivity (law of Least Efforts)
- Success-breeds success phenomenon
- Circulation theory
- Information Product and Processes (IPP) and Duality in IPPs

There are other theoretical studies, especially in the area of circulation theory, citation analysis, sources-items relation, etc. Some of these are discussed in:

- Quantitative Method for Library and Information

Science by I K Ravichandra Rao. Wiley-Eastern. New Delhi. 1983.

- Proceedings of the ISSI Conferences in Scientometrics and Informetrics (held bi-annually since 1987)
- Introduction to Informetrics: Quantitative Methods in Library, Documentation and Information Science. By Leo Egghe and Ronald Rousseau. Elsevier. Amsterdam. 1990
- Power Laws in the Information Production Process: Lotkaian Informetrics. By Leo Egghe. Elsevier. Amsterdam. 2005.

Studies based on citation data have several limitations. They are:

- Citation studies are mostly dependent on SCI data – which cover only a limited number of journals and its coverage does not remain constant since new journals are added regularly and some are dropped,
- Cited articles appear only under the name of the first author, and
- Solutions to problems of eliminating self-citation are cumbersome.

If explanations for library or science of information phenomenon are based solely upon individuals' experiences, then they may not be accurate and /or reliable; they are not amenable to generalization. Thus, a primary objective of bibliometric research is the development of a general and systematic set of theories from which hypotheses can be generated and tested. Scientometric studies vary from each other from several points of view. They adopt different methods of data collection as well as different techniques. Even there are no universally accepted terminologies. In addition, use of algebraic symbols varies from one study to another.

Under these circumstances, it would be difficult to think of "bibliometric or scientometrics standards," let alone formulating them.

Most of the scientometric studies are empirical in nature. In such circumstances, to reproduce the research, one has to repeat the survey and analyze the data right from the beginning. Even then, we may not get the same result! In natural sciences, it is possible and quite common that research may be repeated in laboratories. But in social sciences this is not only difficult, but is not possible. Further, an important cause of the overall unreliability and therefore a cause of invalidity in any basic research in the social sciences are due to space and time factors. It is therefore difficult to reproduce the results of research.

Under the circumstances, if the study is based on certain guidelines/ methodologies, one may accept its validity as well as its generalization of the result. Also, if scientific methods are followed, we may satisfy ourselves with certain statistical parameters — measures of central tendency, measures of dispersion, confidence limits, etc. the general guidelines are:

1. Identify the general problem(s)
2. Conduct literature search
3. Decide the design methodology
4. Collect the data either for the population or for a sample
5. Analyze the data
6. Report the result
7. Refine the hypotheses

(Note Steps 2 to 7 may have to be repeated)

In Step 1, the research objectives are explicitly identified and described. Further, information about the research objectives and investigative tasks are analyzed and relevant terms and variables are defined. The research question is stated and/or hypotheses are formulated in Step 3 with clear-cut definitions, assumptions, suppositions, etc. The objectives of the study are then carefully observed and, if necessary, causal factors associated with the observed phenomenon are identified. If sufficient data are collected from machine-readable databases, results are much more reliable than otherwise. Thus whenever possible, one may use, databases for collecting the data. In most of the studies, several assumptions and /or suppositions are made. The validity of the results often depending on how far such assumptions and suppositions are valid. For instance, Bradford[12] in his study assumed that $n_1 = n_2 = n$, where $n_1 = r/r_1$ and $n_2 = r_1 / r_2$; r, r_1, r_2 are the average numbers of articles per periodical in three different zones. The number of periodicals in the three zones are p, p_1 and p_2. Precisely he mentioned "…. We know no reason why n_1 and n_2 should differ and the simplest supposition we can make is that they are equal. Consequently, in the simplest possible case, we get $p{:}p_1{:}p_2 :: 1{:}n{:}n^2$ "(Bradford, S.C. (1948). "Documentation." Crosby Lockwood, London, p.111) For his data, in the area of applied physics, $n_1 = 3.83$ and $n_2 = 7.26$. For the data in the area of lubrication, $n_1 = 3.41$ and $n_2 = 3.22$. Ravichandra Rao[92] in his study on Methodological and Conceptual Questions of Bibliometric Standards, analyzed a bibliography (Bibliography and Index to Geology. Published by American Geological Institute, Monthly, 1985) consisting of 420 references in 20 journals. It was found that $n_1 = 1.33$ and $n_2 = 3.25$. Also in another study, in the area of economics, he found that $n_1 = 2.38$ and $n_2 = 6.04$. In most of such similar studies, it may be noticed that n_1 and n_2 differ quite significantly. In such cases, how far Bradford's assumption is correct? And thus how far his statement is valid?

In order to arrive at valid results, it is recommended that in every study one must strictly follow the scientific methods as enunciated in the above said general guidelines. A mere presentation of the summary of the data should not be considered as the end results of the research.. If these guidelines are followed, they are as good as bibliometric standards. These are of course methodological and conceptual standards. In addition to the above guidelines, depending on the problems, the following techniques and procedures may be adopted for processing the data:

- Rank and size frequency analyses
- Correlation and regression analyses
- Cluster and factor analyses
- Use of OR techniques
- Statistical inferences, etc.

Thus, what is most important is that there is no single method, which is applicable for all bibliometric research. There are different methods and procedures for different problems and they are different standards. It is evident that bibliometric and/or scientometrics studies are still in empirical stages. The facts are gathered either through surveys or from published bibliographies, indexes, and databases. Based on these facts, empirical models and principles are developed. The normative principles and standards have to evolve from the logical analyses of the empirical models. The stage is set to integrate empirical models of scientometrics into standards. Future, scientometrics studies have to address this issue and reach the stage of normative principles.

Thus in a sense there is a crisis. According to Glanzel and Schoepflin[44], the causes of the crisis are:

- Lack of leadership in scientometrics

- Audience of scientometrics — : different target groups
- Limited scope of Lotka's law and Bradford's law (— based on static data)
- High cost of bibliographic data
- Publication policies and retrieval possibilities offered by the vendors which tend to put limitations on bibliometrics research

Some of the problems are trivial in nature and most of these issues should not be a major concern for scientometrics research. However, to overcome these problems, I suggest that

- To reach a minimum consensus – in relation to theory and terminology
- Field must include all quantitative aspects of models of science communications and retrieval of scientific information.
- Needs interactions with other disciplines
- Need for national research programmes

It is further important to note that research in bibliometrics / informetrics / scientometrics is moving mostly towards theoretical work. In order to appreciate it one must apply his/her research work! One has to put much effort to apply it to the day-to-day activities in relation to "information handling;" it must be applied in libraries or information centers or in science administration. Measures must be identified or developed, not simply for the sake of developing or identifying. Practising professionals must use it – to improve the existing systems and services, to study and understand the systems to reduce their operational costs, etc. Further, we may notice that a gap between the theoretical work in the metrics (— whether scientometrics or bibliometrics or informetrics) and its applications is increasing. Today, econometrics is a well-defined discipline and a key subject, not because of its simple theoretical work but because of its applications and its usefulness to the society. Why not in a similar way, scientometrics?

There is need to consider scientometrics and / or informetrics as one of the core subjects in library and information education programmes. Unless it is taught at the Master's degree level, its importance will not be recognized, students are unlikely to take up this topic in their research activities.

REFERENCES

1. A bibliography on "optical flow estimation." URL: *http://itb1.biologie.hu-erlin.de/~wiskott/Bibliographies/FlowEstimation.htma=ml*
2. Abraham, R.H. (1996). "Webometry: Measuring the complexity of the Complexity of the World Wide Web." (http://www.ralph-abraham.org/vita/redwood/vienna.html)
3. Almind, T.C., and Ingwersen, P. (1997). "Informetric analyses on the World Wide Web: methodological approaches to Webometrics. Toward a basic framework for Webometrics."*Journal of the American Society for Information Science and Technology*, 55 (14), 1216-27.
4. Archibald, G. and Line, M.B. (1990). "The size and growth of serial literature 1950-1987, in terms of the number of articles per serial." *Scientometrics*. 20, 173-196.
5. *Aslib Proceedings* (1948). 1, 102.
6. Batista, Pablo, D and others (2006). "Is it possible to compare researchers with different scientific interests?" *Scientometrics*. 68(1),179-89.
7. Beaver, Donald Deb. (2004). "Does collaborative research have greater epistemic authority?" *Scientometric. 60(3), 399-408.*

8. Bhattacharya, Sujit. and Meyer, Martin. (2004). "Commonalities and differences between scholarly and technical collaboration: an exploration of *co-invention* and *co-authorship* analyses." *Scientometrics.* 61(3), 443-456.
9. Bookstein, A. (1977). "Patterns of Scientific Productivity and Social Change: A discussion of Lotka's Law and Bibliometric Symmetry". *Journal of the American Society for Information Science.* 28(4), 206-210.
10. Bornmann, Lutz and Daniel, Hans-Dieter. (2006). "Does the h-index for ranking of scientists really work?" *Scientometrics.* 65(3), 91-2.
11. Bossy, M.J. (1995). "The last of the litter: Netometrics" http:// biblio.fr.info.unicean.fr/bnum/jelec/Solaris/d02/2bossy.html
12. Bradford S.C. (1934). "Source of information on specific subjects." *Engineering.* 26, Jan., 85-86
13. Brookes, B C. (1970). "Obsolescence of special library periodicals: Sampling Errors and Utility contours" *Journal of the American Society for Information Science.* 21, 320-29
14. Brookes, B.C. (1990). "Biblio-, Sciento-, Infor-metrics??? What are we talking about?" **In** L. Egghe and R. Rousseau (*Eds*) *Informetrics 89/90.* 31-43.
15. Braun, Tibor, Glanzel, Wolfgang and Schubert, Andras. "A Hirsch type index for journals". *The Scientist.* 19(22, 8). Also in *http://www.the_*scientist.com/article/display/15890/
16. Burrell, Q. and Rousseau, R. (1995). "Fractional counts for authorship attribution: a numerical study." *Journal of the American Society for Information Science.* 46, 297-302.

17. Burton R. E. Keblers. and R. W. (1960). "The half-life of some scientific and technical literature". *American Documentation.* 11, 18-22.

18. Chakrabarthi, S., Joshi, M.M., Punera, K., and Pennock, D.M. (2002). The Structure of broad topics on the Web. *Proceedings of the WWW2002 Conference.*

19. Cherno Gorenko, V.B and Muchnik, S.V (1981). "Scientometric Estimation of a present day study on phosphides." *Scientometrics;* 3,(2), 127-134.

20. Coile, Russell C.(1976.) "Bibliometric Studies of Scientific Productivity". *Proceedings of the American Society for Information Science.* 13, 90.

21. Coile, Russell C. (1977.) Lotka's Frequency Distribution of Scientific Productivity. *Journal of the American Society for Information Science.* 28(6), 366-70.

22. Crane, D. (1972.) "Invisible colleges: diffusion of knowledge in scientific communities". The University of Chicago Press. Chicago.

23. Croxton, F.E. and Cowden, D.J. (1966). "Applied general statistics". Prentice-Hall of India (Private) Ltd. New Delhi.

24. Dhyani, D., Keong, W., and Bhowmick, S.S (2002). "A survey of web metrics." (ACM *Computing Survey.* 34(4), 469-503.

25. Donella H. Meadows, Dennis L. Meadows, Jorgen Randers, and William W, Behrens III. (1972). The Limits to Growth". Club of Rome.

26. Egghe, L. (1993). "On the Influence of Growth on Obsolescence". *Scientometrics.* 27(2), 195-214.

27. Egghe, L. (2000). "Heuristic study of the first-citation distribution". *Scientometrics.* 48(3), 345-59.

28. Egghe, L. (2005). "An explanation of disproportionate growth using linear 3-dimensional informetrics and its relation with the fractal dimension." *Scientometrics.* 63(2), 277-96.
29. Egghe, L. (1993). "Consequence of Lotka's law in the case of Fractional counting of authorship and of Author Count". *Math Computing Modeling.* 18(9), 63-77.
30. Egghe, L. (2005). "Power Laws in The Information Production Process" *Lotkaian Informetrics.* Elsevier.
31. Egghe, L. (2006). "Theory and practice of the g-index." *Scientometrics.* 69(1), 131-52.
32. Egghe, L. and Rousseau, R. (2006). "An informetric model for the Hirsch-index." *Scientometrics.* 69(1), 121-29.
33. Egghe, L. "Distributions of the h-index and the g-index" (to be published.)
34. Egghe, L and Rao, I.K.R. (2002). "Duality Revisited: Construction Fractional Frequency Distributions based on two dual Lotka Laws". *Journal of the American Society for Information Science and Technology.* 53(10), 789-801.
35. Egghe, L. and Rao, I.K.R. (1992). "Classification of growth models based on growth rates and its applications." *Scientometrics.* 25, 5-46.
36. Egghe, L and Ravichandra Rao, I.K. (1992). "Citation age data and the obsolescence function: fits and explanations." *Information Processing and Management.* 28(2), 201-17.
37. Egghe, L.; Ravichandra Rao, I.K. and Rousseau, R. (1995). "On the influence of production on utilization functions: Obsolescence or increased use?" *Scientometrics.* 34(2), 285-315.

38. Egghe, L and Ravichandra Rao, I.K. (2001). "Theory of first citation distributions and applications" *Mathematical and Computer Modeling.* 34, 81-90.

39. Egghe, L and Ravichandra Rao, I.K. (2002). "Theory and Experimentation on the most recent reference Distribution". *Scientometrics.* 53(3), 371-87.

40. Egghe, L. and Rousseau, R. (1990). "Introduction to informetrics". Elsevier. Amsterdam.

41. Egghe, L, Rousseau, R. and Van Hooydonk, G. "Method for accrediting publications to authors or countries: consequences for evaluation studies." *Journal of the American Society for Information Science.* 51(2), 145-57

42. Exponential Model. <http://www.ento.vt.edu/~sharov/ PopEcol/lec5/exp.html>.

43. Fairthorne, R.A. (1969). "Empirical Hyperbolic Distributions (Bradford-Zipf-Mandelbrot) for bibliometric description and prediction." *Journal of Documentation.* 25, 319-43.

44. Glanzel, w and Schoepflin, U. (1994). "Little Scientometrics-Big Scientometrics. and beyond." *Scientometrics.* 30(2/3), 375-84.

45. Gompertz, B. (1825). "On the Nature of the Function Expressive of the Law of Human Mortality, and on a New Mode of Determining the Value of Life Contingencies." *Philosophical Transactions of the Royal Society of London.* 115, 513-85.

46. Griffith, Belver C and others. Informal Contacts in Science: a probabilistic model for communication processes. *Science.* 173(164-5), 1971.

47. Gross, P. L. K. and Gross, E. M. (1927). "College Libraries and chemical education". *Science.* 66, 1229-34.

48. Guan, J.C. and Ma, N. (2004). "Research performance in Computer Science." *Scientometrics.* 61(3), 339-59
49. Gupta, B.M. (1998). "Growth and obsolescence of literature in theoretical population genetics". *Scientometrics.* 42(3), 335-47.
50. Gupta, B.M.; Sharma, Praveen and Suresh Kumar. (1999). "Growth of world and Indian literature". *Scientometrics.* 44, 5-46.
51. Gupta, B.M. and Karisiddappa, C.R. (2000). "Modeling the growth of literature in the area of theoretical population genetics." *Scientometrics.* 49(2), 321-55.
52. Gupta, B.M., *et al.* (2002). "Modeling the growth of world social science literature." *Scientometrics.* 53, 161-64.
53. Hartley, James. (2005). "Refereeing and the single author." (*Journal of Information Science.* 31(3), 251-256.
54. Hersh, A. H. (1942). "Drosophilae and the course of research." *Ohio Journal of Science.* 42(5), 198-200.
55. Hirsch, J. E. "An index to quantify an individual's scientific research output." *http://www.pnas.org/cgi/content/abstract/102/46/16569. http://arxiv.org/abs/physics/?0508025*
56. Hulme, E.W. (1923). "Statistical Bibliography in relation to the Growth of Modern Civilization." Grafton. London.
57. Ingwersen, P. (1998). "The calculation of web impact factors." *Journal of Documentation,* 54(2), 236-43.
58. Ivanisevic, Ranka. and Sapunar, Damir. (2006). "Multiple *authorship* in a small medical journal: a case study of the Croatian Medical Journal". *Journal*

of the American Society for Information Science and Technology, 57(8), 1073-78.

59. Jing, P. and Kang, Z. (2000). "On the mathematical models of the growth of scientific literature". *Journal of the China Society for Scientific and Technical Information*. 19(1), 90-6.
60. Kim, Ki-Wan. (2006). "Measuring international research collaboration of peripheral countries: taking the context into consideration". *Scientometrics*. 6(2), 231-40.
61. Kretschmer, Hildrun; Kretschmer, Ute; Kretschmer, Theo. (2007). "Reflection of *co-authorship* networks in the Web: Web hyperlinks versus Web visibility rates". *Scientometrics*. 70(2), 519-40.
62. Kundra, Ramesh. (2002). "Quantitative Analysis of Medical Literature in India: A Bibliometric Study". *Thesis submitted to the Dept. of Library and Information Science of the Pune University for the Award of Degree of Doctor of Philosophy in Library and Information*. Guide: I.K.Ravichandra Rao, DRTC, Indian Statistical Institute, Bangalore.
63. Lal, Arjun and Ray, P.K. (1993). "Pattern of Research Contribution in leading Horticulture Journal of the World: a Comparative study". *IASLIC Bulletin*. 36(3), 95-102
64. Li-chun, Yin et al. (2006). "Connection and stratification in research collaboration: an *analysis* of the COLLNET network." *Information Processing & Management*. 42(6), 1599-1613.
65. Lindsey, D. (1980). "Production and Citation Measures in the Society of Science: The Problem of Multiple Authorship". *Social Studies of Science*. 10, 145-62.

66. Line, Maurice B. and Sandison, A. (1974). "Obsolescence and change in the use of literature with time" *Journal of Documentation.* 30(3), 283-350.
67. Lotka, A.J. (1926). "The Frequency Distribution of Scientific Productivity". *Journal of the Washington Academy of Science.* 16(12), 317-23.
68. Logistic Functions. <http://www.wmueller.com/precalculus/families/1_81.html>.
69. Lundberg, Jonas *et al* (2006). "Collaboration uncovered: exploring the adequacy of measuring university-industry collaboration through co-authorship and funding." *Scientometrics.* 69(3), 575-89.
70. Mabe, Michael. (2003). "The growth and number of journals." *Serials.*16(2), 191-97.
71. Maheshwarappa, B.S. and Ningoji, M.M. (1993). "Growth of Literature in the field of Applied Sciences in India (1965-89)". *International Information Communication and Education.* 12(2), 1993, 191-200.
72. Malthus, Thomas Robert. (1798). "An essay on the Principle of Population: as it affects the Future Improvement of Society", with Remarks on the Speculations of Mr. Godwin, M. Condorcet, and other Writers. London: J. Johnson, in St. Paul's Church Yard, 1798, 1st Ed.
73. Matia, Kaushik, *et al.* (2005). "Scaling phenomena in the growth dynamics of scientific output." *Journal of the American Society for Information Science and Technology.* 56(9), 893-902.
74. Mendez, A.; Fernandez, M.F. and Lopez, Aguada G. (1987). "Six Year Spanish Scientific Activity in Physics and Engineering through INSPEC and COMPENDEX". *Scientometrics.* 12(1/2), 81-100.

75. May, K.O. (1966). "Quantitative growth of the mathematical literature". *Science.* 154, 1672-73.
76. Menard, H.W. (1974). "Science: growth and change." Harvard University Press, Cambridge, Mass.
77. Murphy, L.J. (1973). "Lotka's Law in Humanities?" (*Journal of the American Society for Information Science. 24(6), 161-162.*
78. Naranan, S. (1970). "Bradford's law of bibliography of science: an interpretation." *Nature.* 227(5258), 631-32.
79. Narin Francis. "Evaluative Bibliometrics: the use of publication and citation analysis in the evaluation of scientific activity." Computer Horizons, Inc. Project No. 704R.
80. Neelameghan A. (1963). Documentation of the History of Medicine in India. *Annals of Library Science and Documentation.* 10(3/4), 116-42.
81. Noruji, A. (2006). "Web presence and Impact Factors for Middle-Eastern Countries." *Online.* 30(2), 22-28.
82. Parvathamma, N.; Gunjal, S.R. and Nijagunappa, R. (1993). "Growth pattern of literature and scientific productivity of authors in Indian earth science (1978-88): a bibliometric study". *Library Science with a Slant to Documentation.* 30(2), 54-64.
83. Persson, O.; Glänzel, W. and Danell, R. (2003). "Inflationary bibliometric values: the role of scientific collaboration and the need for relative indicators in evaluative studies." *Proceedings of the Ninth International Conference on Scientometrics and Informetrics, Beijing (China),* Jiang Guohua, R. Rousseau and Wu Yishan, eds.,411-420, Dalian University of Technology Press, Dalian (China).
84. Price Derek De Solla. (1963). "Little Science, Big

Science". Columbia University Press. New York (GS201).

85. Price Derek De Solla. (1971). "Some remarks on elitism in information and the invisible college phenomenon in science". *Journal of the American Society for Information Science.* 22, 74-75.
86. Price, Derek De Solla. (1976). "A General Theory of Bibliometric and other Cumulative Advantage Processes". *Journal of the American Society for Information Science.* 27(5), 292-306.
87. Pritchard, A. (1969). "Statistical Bibliography or Bibliometrics". *Journal of Documentation.* 25(4), 348-9.
88. Rajan, T.N. and Sen, B.K. (1985). "An Essay on Informetrics: A Study in Growth and Development". **In** *Proceedings of the IASLIC XV,* Bangalore.
89. Ramakrishna, N.V. and Pangannaya, N.B. (1999). "Growth of animal cell culture technology literature: a correlation between citations and publications based on growth curves". *Library Science with a Slant to Documentation and Information Studies.* 36(1), 21-26.
90. Ravichandra Rao, I.K. (1981). "Distribution of Scientific productivity and social change". *Journal of American Society for Information Science.* 31(2), 111-21.
91. Ravichandra Rao, I.K. (1985). "Informetrics vis-à-vis Bibliometrics". *FID/IM Newsletter.* No. 1, June.
92. Ravichandra Rao, I.K. (1995). "A stochastic Approach to Analysis of Distribution of Papers in Mathematics: Lotka's Law Revisited". *Proceedings of the Fifth Biennial Conference of the International Society for Scientometrics and Informetrics.* Royal College, River Forest, IL, 455-64.

93. Ravichandra Rao, I.K. (1996). Methodological and Conceptual Questions of Bibliometric Standards. *Scientometrics.* 35(2), 265-70.

94. Ravichandra Rao, I.K. and Meera, B.M. (1991). "Growth and obsolescence of literature: an empirical study." **In** I.K.R. Rao ed. *Informetrics – 91.* Sarada Ranganatahan Endowment for library. Bangalore. 377-394.

95. Ravichandra Rao, I.K.; Sahoo, Bibhuti Bhusan and Egghe, L. (2003). "A Distribution of papers based on Fractional Counting: An Empirical Study." *Proceedings of the 9th International Conference on Scientometrics and Informetrics.* Dalian University of Technology Press. Dalian, China. 241-250.

96. Ravichandra Rao, I.K. and Sahoo, Bibhuti Bhushan. Distribution of Multiple Authors: A Case Study of two journals (*JASIST and Scientometrics and in the subject software studies*) (*Proceedings of the Fourth International Conference on Webometrics, Informerics and Scientometrics (WIS)*, 28th July – 1st August 2008, Berlin).

97. Ravichandra Rao, I.K. and Sahoo, Bibhuti Bhushan. (2008). Distribution of Multiple Authors: A Case Study of two journals (JASIST and Scientometrics) *COLLNET Journal of Scientometrics and Information Management.* 2(1), 27-36.

98. Ravichandra Rao, I.K. (2007). "Distributions of Hirsch-index and g-index: An Empirical Study *Proceedings of ISSI-2007,* CSIC, Madrid, Spain. June 25-27, V II, 655-58.

99. Ravichandra Rao, I.K. and Divya Srivastava. "Growth of Journals, Articles and Authors". (accepted for publication in Journal of Informetrics)

100. Rousseau, R. (1992). "Breakdown of the Robustness

of Lotka's Law: The case of Adjusted Counts for Multi-authorship Attribution". *Journal of the American Society for Information Science.* 43(10), 645-47.

101. Rousseau, R. Simple models and the corresponding h- and g-index. (Submitted to Science and Technology Development.)

102. Sahoo, Bibhuti Bhusan. (2006). "Scientometric Study of Literature in Software Studies in India with a comparison to the World Literature." *Thesis Submitted to the Dept. of Library and Information Science of the Pune University for the Degree of Doctor of Philosophy in Library and Information Science.* Guide: I. K. Ravichandra Rao, Documentation Research and Training Center, Indian Statistical Institute, Bangalore.

103. Schockley, William. (1957). "On the statistics of individual variation of productivity in research laboratories". *Proceedings of IPE.* 279-90.

104. Schorr, Alan Edward. (1995). "Lotka's law and Map Librarianship." *Journal of American Society of Library and Information Science.* 26(3), 189-90.

105. Sharma, Praveen; Gupta, B.M. and Suresh Kumar. (2002). "Application of selected growth models to growth of science and technology literature in research specialties". *DESIDOC Bulletin of Information Technology,* 22(2), 17-25.

106. Simon, H. A. (1955). "On a class of Skew Distribution Functions *Bibliometrika.* 42, 425-40.

107. Suresh Kumar and Garg, K.C. (2005). "Scientometrics of Computer Science Research in India and China." *Scientometrics.* 64(2), 121-132.

108. Tague, J.; Beheshti, J. and Rees-Potter, L. (Summer 1981). "The law of exponential growth: evidences,

implications and forecasts." *Library Trends.* 125-49.

109. Tsay, Ming-Yueh and Yang, Yen-Hsu. (2005). "Bibliometric analysis of the literature of randomized controlled trials." *Journal of the Medical Library Association (JMLA)* 93(4).

110. Tsay, Ming-Yueh and Yang, Yen-Hsu. (2003): "A study on the literature growth and author productivity of randomized controlled trials medical literature." *Journal of Information, Communication and Library Science 9.3* (chased 4/6/2004 Mar.2003). 31-46.

111. van Raan, Anthony, F. J. "Comparison of the Hirsch-type index with standard bibliometric indicators and with peer judgment for 147 research groups". (http://arXiv:physics/0511206 Nov 2005). http://adsabs.harvard.edu/abs/2005physics.11206V

112. Vijay, K.R. (2005). "Bibliometric Study of Research Publication Trends among Indian Food Scientists and Technologists." *Annals of Library and Information Studies.* 52(3), 77-81.

113. Wolfram, D.; Chu, C.M. and Liu, Xin. (1990). "Growth of knowledge: Bibliometric analysis using online databases data in L. Egghe and R. Rousseau, eds. *Informetrics 89/90.* 355-72.

114. Williams, C.B. (1944). "Number of publication written by biologists". *Annals of Human Genetics.* 12, 143-46.

115. Pravdic, Nevenka and Oluic-Vukovic, Vesna. (1991). "Distribution of Scientific Productivity: Ambiguities in the Assignment of Author Rank." *Scientometrics.* 20(1), 131-44.

INDEX